Introduction to Bacteria
FOR STUDENTS OF
BIOLOGY, BIOTECHNOLOGY & MEDICINE

Introduction to Bacteria

FOR STUDENTS OF BIOLOGY, BIOTECHNOLOGY & MEDICINE

SECOND EDITION

Paul Singleton

JOHN WILEY & SONS

Chichester · New York · Brisbane · Toronto · Singapore

1st edition 1981
Reprinted April 1985
Reprinted February 1987
Reprinted July 1989
Reprinted April 1991

Japanese edition 1982
French edition 1984

2nd edition 1992

Other Wiley Editorial Offices

John Wiley & Sons, Inc., 605 Third Avenue,
New York, NY 10158-0012, USA

Jacaranda Wiley Ltd, G.P.O. Box 859, Brisbane,
Queensland 4001, Australia

John Wiley & Sons (Canada) Ltd, 22 Worcester Road,
Rexdale, Ontario M9W 1L1, Canada

John Wiley & Sons (SEA) Pte Ltd, 37 Jalan Pemimpin #05-04,
Block B, Union Industrial Building, Singapore 2057

Library of Congress Cataloging-in-Publication Data

Singleton, Paul.
    Introduction to bacteria : for students of biology, biotechnology,
and medicine/Paul Singleton.—2nd ed.
        p.   cm.
    Includes bibliographical references and index.
    ISBN 0 471 93146 2 (paper)
    1. Bacteria.   I. Title.
QR75.S56 1992
589.9—dc20   92-8941
                                                              CIP

British Library Cataloguing in Publication Data

A catalogue record for this book is
available from the British Library

ISBN 0 471 93146 2

Typeset in 10/12pt Palatino by GCS, Leighton Buzzard
Printed and bound in Great Britain by Biddles Ltd, Guildford.

Contents

Preface

Like the first edition, this book assumes no prior knowledge of bacteria. Starting at the beginning, it builds an integrated picture of the structure, growth, differentiation, metabolism and molecular biology of the bacteria; these 'core' aspects are central to *all* areas of bacteriology—including biotechnology, medicine, food technology, water supplies, sewage treatment, agriculture and environmental science. Subsequent chapters deal with specific areas of applied bacteriology and also cover the classification/ identification of bacteria, disinfection, sterilization and other areas of practical bacteriology.

Since the first edition there have been many advances, both in concept and detail; these advances have shaped the second edition, and, in general, the subject has been taken to a higher level. In the text, however, knowledge is built up sequentially, from first principles, so that students may either study at an elementary level or progress to a level at which they can follow more advanced texts and papers; for the latter group, the text includes a number of references to recent papers and reviews.

Acknowledgements. For some of the material I am indebted to an extraordinarily fine scientist and writer: Diana Sainsbury, co-author of the first edition of this book and of two editions of *Dictionary of Microbiology and Molecular Biology.* I also thank those fellow scientists who were generous with papers, photographs and time, and I am most grateful to Clive Ward, Medical Librarian, for the use of his excellent library at the University of Bristol. Finally, I thank the Wiley team at Chichester for publishing the book ahead of schedule.

1 The bacteria: an introduction

1.1 WHAT ARE BACTERIA?

Bacteria are minute organisms which occur almost everywhere. They sometimes reveal their presence—wounds 'go septic', milk 'sours', meat 'putrefies'—but usually we are unaware of them because their activities are less obvious and because they are so small. Indeed, the very existence of bacteria was unknown until the development of the microscope in the 17th century.

In most cases a bacterium is a single, autonomous cell. However, the bacterial cell is unlike the cells of other organisms. Bacterial cells are of the *prokaryotic* type, while those of all other organisms (including higher animals and plants) are of the *eukaryotic* type. Prokaryotic and eukaryotic cells differ in many important ways. For example, in a eukaryotic cell the chromosomes (the thread-like bodies of genetic material) are enclosed in a membranous sac to form a distinct structure, the nucleus; in prokaryotic cells there is no nuclear membrane—the chromosome(s) being in direct contact with the cell's cytoplasm. The main differences between prokaryotic and eukaryotic cells are summarized in Table 1.1; prokaryotic structures mentioned in the table will be considered in more detail in later chapters.

Bacteria are included in the category 'microorganisms'. The microorganisms include several distinct types of organism—algae, fungi, lichens, protozoa, viruses and subviral agents—as well as bacteria; hence, though all bacteria are microorganisms, not all microorganisms are bacteria.

Fossil organisms resembling some of today's *cyanobacteria* (see later chapters) are among the oldest fossils known. From fossil records and other data it seems that cyanobacteria may have been the dominant form of life about 2500–570 million years ago—in the middle to late Pre-Cambrian; in fact, they were probably responsible for generating the Earth's atmospheric oxygen.

1.2 WHY STUDY BACTERIA?

One important reason is the conquest of disease. Bacteria cause some major diseases as well as a number of minor ones; the prevention and control of these diseases depend largely on the efforts of medical, veterinary and agricultural bacteriologists. Pathogenic (i.e. disease-causing) bacteria are considered in Chapter 11.

Table 1.1   Eukaryotic and prokaryotic cells: some major differences

Eukaryotic cells	Prokaryotic cells
The chromosomes are enclosed within a sac-like, double-layered 'nuclear' membrane	There is no nuclear membrane: chromosomes are in direct contact with the cytoplasm
Chromosome structure is complex; the DNA is usually associated with proteins called histones	Chromosome structure is relatively simple; histone-like proteins have recently been discovered in some bacteria
Cell division involves mitosis or meiosis	Mitosis and meiosis are not involved
The cell wall, when present, includes structural compounds such as cellulose or chitin, but never peptidoglycan	The cell wall, when present, usually contains peptidoglycan, but never cellulose or chitin
Mitochondria are generally present; chloroplasts occur in photosynthetic cells	Mitochondria and chloroplasts are never present
Cells contain ribosomes of two types: a larger type in the cytoplasm and a smaller type in mitochondria and chloroplasts	Cells contain ribosomes of only one size
Flagella, when present, have a complex structure	Flagella, when present, have a relatively simple structure

Important though they are, the pathogenic bacteria are only a small proportion of the bacteria as a whole. Most bacteria do little or no harm, and many are positively useful to man. Some, for example, produce antibiotics which have revolutionized the treatment of disease, while others provide enzymes for 'biological' washing powders. Some are used as 'microbial insecticides'—protecting crops from certain insect pests. Bacteria are even used to leach out metals from some low-grade ores. Perhaps surprisingly, bacteria contribute a lot to the food industry (Chapter 12). We usually think of bacteria as a nuisance where food is concerned, causing 'spoilage' and 'food poisoning', but certain types are actually employed in food production. For example, in the manufacture of butter, cheese and yoghurt, certain bacteria are used to convert 'milk sugar' (lactose) to lactic acid; the bacteria also form compounds which give these products their characteristic flavours. Vinegar is produced from alcohol (ethanol) by bacterial action, and bacteria also play a part in the manufacture of cocoa and coffee. Some of these activities can be understood by studying the chemical reactions (metabolism) of bacterial cells (Chapters 5 and 6). Chapters 12 and 13 look at the activities of some 'useful' bacteria.

Not least, bacteria have essential roles in the natural cycles of matter

(Chapter 10)—on which, ultimately, all life depends. In the soil, bacteria affect fertility and structure—agricultural potential—so that a better understanding of bacterial activity will permit better management of land and crops; in the future this will be vital to the survival of our ever-expanding population.

From this brief summary it should be clear that the more we learn about bacteria the more effectively we can minimize their harmful potential and exploit their useful activities.

1.3   CLASSIFYING AND NAMING BACTERIA

How is one type of bacterium distinguished from another, and how are bacteria classified? Bacteria may differ, for example, in their shape, size and structure, in their chemical activities, in the types of nutrients they need, in the form of energy they use, in the physical conditions under which they can grow, and in their reactions to certain dyes.

Features such as those listed above are widely used for classifying (and identifying) bacteria—such features being easily checked even in a modestly equipped laboratory. As in other areas of biology, bacteria are classified in a hierarchy of categories—e.g., families, genera, species; *species* which are sufficiently alike are placed in the same *genus*, and genera with a certain level of similarity are grouped into a *family*. A species may be subdivided into two or more *strains*—organisms which conform to the same species definition but which have minor differences. In general, members of (say) a bacterial family would have similar structure, would use the same form of energy, and would typically react in a similar way to certain dyes; the species in such a family may be grouped into genera on the basis of differences in chemical activities, nutrient requirements, conditions for growth and (to some extent) shape and size.

Although useful for everyday purposes, the kind of classification described above does not necessarily indicate *evolutionary* relationships among the bacteria. In the last decade or so, studies have been carried out on what are believed to be more fundamental characteristics of bacteria; such characteristics are believed to reveal major evolutionary pathways. This aspect of classification is referred to in later chapters.

As in the case of animals and plants, each species of bacterium is given a name in the form of a Latin binomial. A binomial consists of (i) the name of the genus to which a given organism belongs, followed by (ii) the 'specific epithet' which acts as a label for one particular species; for example, *Escherichia coli* gives the name of the genus (*Escherichia*) and the specific epithet (*coli*). By convention, a Latin binomial is printed in italics, or is underlined once if handwritten; the name of a genus always begins with a capital letter, but a specific epithet always begins with a lower-case letter.

The name of a species may be abbreviated by abbreviating the name of the genus—e.g. *'Escherichia coli'* may be written *'E. coli'*; however, this should be done only when the full name of the genus has been mentioned earlier so that the meaning of the abbreviation is clear. (Being a very common experimental organism, *E. coli* is mentioned frequently in this book; the name—and spelling—of the genus should be noted.)

The names of families and orders of bacteria are not printed in italics, but each has a capital initial letter. These names also have standardized endings, the name of a family always ending in '-aceae' (e.g. Enterobacteriaceae) and the name of an order always ending in '-ales' (e.g. Actinomycetales).

The naming of bacteria is formally governed by various rules made by the International Committee on Systematic Bacteriology. The advantages of an internationally standardized system of naming are obvious, but the rules are not always adhered to in the literature—owing to a lack of awareness (of the rules, or of revised names) or to disagreement with published opinions.

2 The bacterial cell

2.1 SHAPES, SIZES AND ARRANGEMENTS OF BACTERIAL CELLS

2.1.1 Shape

Bacterial cells vary widely in shape, according to species. Rounded or 'spherical' cells—of any species—are called *cocci* (singular: *coccus*). Elongated, rod-shaped cells of any species are called *bacilli* (singular: *bacillus*), or simply *rods*. Cocci are not necessarily exactly spherical, and not all bacilli have exactly the same shape; for example, some cocci are more or less kidney-shaped, and some bacilli taper at each end (*fusiform* bacilli) or are curved (*vibrios*). Ovoid cells, intermediate in shape between cocci and bacilli, are called *coccobacilli* (singular: *coccobacillus*). There are also two types of spiral cell: those which are more or less rigid (*spirilla*, singular: *spirillum*) and those which are flexible (*spirochaetes*, singular: *spirochaete*). Then there are the so-called 'square bacteria' (flat, square bacteria) and 'box-like bacteria' (variously-shaped, angular bacteria). Finally, there are the fungus-like *actinomycetes*: bacteria most of which grow in the form of fine threads called *hyphae* (singular: *hypha*); a group or mass of hyphae is referred to as *mycelium*. Bacteria of various shapes are shown in Fig. 2.1 and in Plate 1 (p. 17).

As seen in the caption of Fig. 2.1, some of the names used to describe shapes of bacterial cells are also used as names for bacterial genera. Care should be taken, for example, not to confuse 'bacillus' with '*Bacillus*' (note that the latter has a capital 'B' and is printed in italics); some bacilli belong to the genus *Bacillus*, others do not!

Although the cells of a given species of bacterium are usually more or less uniform in shape, in some species the shape of the cell typically varies from one cell to another—sometimes quite markedly; this phenomenon is called *pleomorphism*.

L-form cells are irregularly-shaped or spherical cells which are produced spontaneously by some species of bacteria, and which can be induced in other species e.g. by temperature shock and by other kinds of physico-chemical stimulus; these cells were named after the Lister Institute of Preventive Medicine (London).

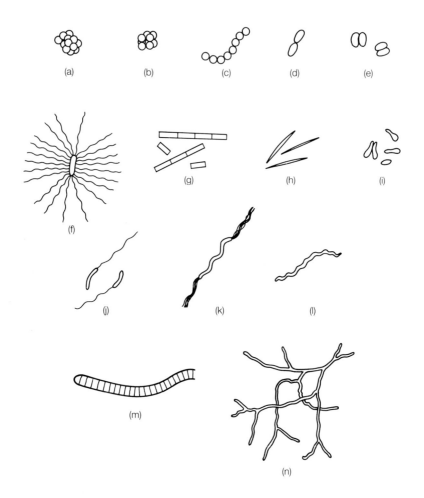

Fig. 2.1 Shapes and arrangements of some bacterial cells with named examples (not drawn to scale). (a) Uniform spherical cells (cocci) in irregular clusters: *Staphylococcus aureus*. (b) Cocci, in regular packets of eight cells: *Sarcina ventriculi*. (c) Cocci in chains: *Streptococcus pyogenes*. (d) Slightly elongated cocci in pairs (diplococci): *Streptococcus pneumoniae*. (e) Pairs of cocci (diplococci) in which each cell is flattened or slightly concave on the side next to its neighbour: *Neisseria gonorrhoeae*. (f) Rod-shaped cell (bacillus): *Escherichia coli*. The lines arising from the bacillus represent fine, hair-like appendages called *flagella* which are described later in the chapter. (g) Blunt-ended bacilli, singly and in chains: *Bacillus anthracis*. (h) Bacilli with tapered ends (fusiform bacilli): *Fusobacterium nucleatum*. (i) Irregularly-shaped (pleomorphic) cells: *Corynebacterium diphtheriae*. (j) Curved bacilli (vibrios), each with one flagellum: *Vibrio cholerae*. (k) A rigid spiral cell (spirillum) with a tuft of flagella at each end: *Spirillum volutans*. (l) A flexible spiral cell (spirochaete): *Treponema pallidum*. (m) One end of a filament (*trichome*) of a cyanobacterium: *Oscillatoria limnetica*. Trichomes are discussed in section 2.3. (n) Thin, branched filaments (hyphae): *Streptomyces albus*.

2.1.2 Size

Bacterial cells are usually measured in *micrometres*, μm (formerly microns, μ); 1 μm = 0.001 mm. Bacteria range in size from about 0.2 μm (e.g. cells of *Chlamydia*) up to about 250 μm (e.g. cells of *Spirochaeta*). However, these are extreme cases; in most species the maximum dimension of a cell lies within the range 1–10 μm. Note that the smallest bacteria are of the same dimensions as the limit of resolution of a good light microscope, which is about 0.2 μm.

2.1.3 Arrangements of bacterial cells

Under the microscope, bacteria of a given species may be seen as separate (individual) cells or as cells in characteristic groupings. According to species, cells may occur in pairs, in irregular clusters, in chains or filaments, in regular *packets* of four, eight or more cells, or in *palisade* form—a number of elongated cells side-by-side in a row with adjacent cells touching. The species *Pelodictyon clathratiforme* is unusual in that it forms three-dimensional networks of cells. In a number of species the cells form *trichomes* (section 2.3.1). Some arrangements of cells are shown in Fig. 2.1. These different arrangements of cells do not result from the aggregation of previously single cells; they occur because (i) cells of the different species divide (reproduce) in different ways, and (ii) two or more cells may remain attached after the process of cell division.

In nature, stable, mixed-species groups of cells called *consortia* are formed by some types of bacteria.

2.2 THE BACTERIAL CELL: A CLOSER LOOK

Are all bacteria basically similar in structure? No: cells of different species may differ greatly both in their fine structure (ultrastructure) and chemical composition; for this reason there is no 'typical' bacterium. Figure 2.2 shows a 'generalized' bacterium in a very diagrammatic way; it is important to note that not all bacteria have all the features shown in the diagram, and that some bacteria have structures not shown in the diagram.

Figure 2.2 shows the cell's *chromosome* ('genetic blueprint')—a loop-like structure of DNA (Chapter 7) which is extensively folded to form a body called the *nucleoid*. Bathing the nucleoid is a complex fluid, the *cytoplasm*, which fills the interior of the cell. The cytoplasm contains *ribosomes*: minute bodies involved in the synthesis of proteins; sometimes there are also *storage granules* of reserve nutrients etc. The nucleoid, cytoplasm, ribosomes and storage granules all occur within the space bounded by a membranous sac, the *cytoplasmic membrane* (also called cell membrane or plasma membrane).

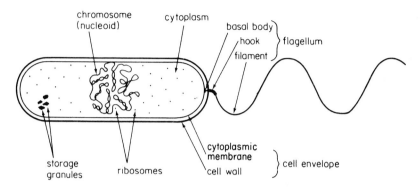

Fig. 2.2 Cross-section of a generalized bacterial cell (diagrammatic).

The outermost layer shown in Fig. 2.2 is a tough (mechanically strong) *cell wall*. Between the cytoplasmic membrane and cell wall is the so-called *periplasmic region*; in some bacteria this region actually includes the inner layer of the cell wall (see later). The cytoplasmic membrane and cell wall are referred to, jointly, as the *cell envelope*. The *flagellum* is a thin, hair-like proteinaceous appendage which is attached by specialized structures to the cell envelope. These and other features of the bacterial cell are considered below.

2.2.1 The nucleoid

The chromosome is essentially a loop of *deoxyribonucleic acid* (DNA) (described in Chapter 7). In at least some bacteria the DNA is associated with proteins resembling the 'histones' in eukaryotic cells. Within the cell, the extensively folded DNA forms a dense body, the nucleoid, which is attached to the cytoplasmic membrane. In the bacterium *Escherichia coli* the chromosome is about 1.3 mm long; all of this DNA fits into a cell of only a few micrometres in length—with room to spare!

   A bacterium may contain more than one copy of the chromosome, according to species and conditions of growth (Chapter 3). In *E. coli*, for example, cells undergoing rapid growth have more chromosomes per cell than do those undergoing slow growth. In some species each cell normally has many copies of the chromosome.

2.2.2 The cytoplasm

The cytoplasm is an aqueous (water-based) fluid containing ribosomes, nutrients, ions, enzymes, waste products and various molecules involved in synthesis, cell maintenance and energy metabolism; storage granules may be present under certain conditions. Bacterial cytoplasm lacks the equivalent of the endoplasmic reticulum in eukaryotic cells; little appears to be known about its organization.

Exceptionally (in a few species), the cytoplasm may contain some unique and interesting items. In *Bacillus thuringiensis*, for example, it sometimes contains crystals that are poisonous for various insects; this species is used in agriculture for biological control (Chapter 13). Other bacteria, themselves parasites of protozoa, contain curious rolled-up ribbons of protein called 'R bodies'. In some aquatic bacteria, tiny particles of magnetite (Fe_3O_4) called *magnetosomes* cause the cells to align in a magnetic field.

2.2.3 Ribosomes

Ribosomes are minute, rounded bodies, each about 0.025 μm, made of RNA (a polymer similar to DNA—Chapter 7) and protein. They are the sites where proteins are synthesized (Chapter 7), and the cytoplasm contains a large number of them (Plate 4: top).

Bacterial ribosomes are smaller than those in the cytoplasm of eukaryotic cells, but they are similar (in size) to those in the chloroplasts of plants and algae. Ribosomes are usually described not by their diameters but by their rate of sedimentation in an ultracentrifuge; measurement is made in Svedberg units (S)—the higher the value the more rapid is the rate of sedimentation. Each bacterial ribosome (70S) consists of one 50S subunit and one 30S subunit; the parts of a ribosome are held together probably by hydrogen bonding and by ionic and hydrophobic interactions—magnesium ions generally being important in maintaining the structure.

Most of a ribosome (about 70% by mass) is RNA (ribosomal RNA, or *rRNA*); a ribosome's function seems to depend primarily on its rRNA. Like the ribosomes themselves, rRNA molecules are also measured in Svedberg units (S). The 30S ribosomal subunit contains 16S rRNA, while the 50S subunit contains 5S and 23S rRNA.

rRNA is a polymer, and it is generally believed that its sequence of subunits (nucleotides) remains relatively unchanged over evolutionary periods of time. Differences in these sequences may therefore indicate evolutionary differences between organisms; for example, in different species of the same genus, 16S rRNA typically differs by at least 1.5%. In recent years, 16S rRNA has been used to classify bacteria into categories that are believed to have evolutionary significance; for example, a difference in 16S rRNA is one of the reasons for distinguishing one major group of bacteria, the Eubacteria, from another, the Archaebacteria.

Ribosomal proteins (r-proteins) account for about 30% (by mass) of a ribosome. In *E. coli*, the 30S subunit has 21 different types of r-protein, while the 50S subunit has over 30 different types.

2.2.4 Storage granules

Under appropriate conditions, many bacteria produce polymers which are stored as granules in the cytoplasm; these compounds include poly-β-hydroxybutyrate and polyphosphate.

2.2.4.1 Poly-β-hydroxybutyrate (PHB)

PHB is a linear polymer of β-hydroxybutyrate (Fig. 2.3). In some species, granules of PHB accumulate when decreasing supplies of nutrients (other than carbon) restrict the cell's rate of growth (Chapter 3); in these cells the PHB acts as a reservoir of carbon and energy (Chapters 5 and 6)—to be used when other nutrients become more plentiful. In the soil bacterium *Azotobacter beijerinckii*, PHB accumulates (up to 80% of the cell's mass) when oxygen is scarce; in this species PHB can 'replace' oxygen as a source of oxidizing power.

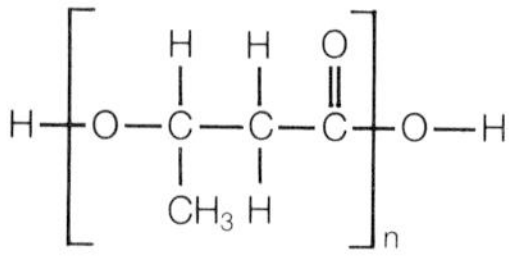

Fig. 2.3   A common storage compound in bacteria: poly-β-hydroxybutyrate.

Each PHB granule is usually bounded by a simple membrane, about 2–4.5 nm thick, to which is attached the enzyme (PHB synthetase) involved in the synthesis of PHB (Chapter 6). PHB granules can be stained, in the cell, by dyes such as Nile blue A.

2.2.4.2 Polyphosphate (polymetaphosphate; volutin)

Polyphosphate ($PO_3^{2-} - O - [PO_3^-]_n - PO_3^{2-}$) granules occur in most types of bacteria. They are believed to act as reservoirs of phosphate and, in some cases, they appear to be involved in energy metabolism; polyphosphate may also serve to store or sequester cations.

When treated with certain dyes (e.g. polychrome methylene blue), polyphosphate granules develop a colour different to that of the dye used to stain them; this phenomenon is called 'metachromacy', and the granules are sometimes called *metachromatic granules*.

2.2.5 Gas vacuoles

Gas vacuoles occur only in certain (typically photosynthetic, aquatic) bacteria. Each vacuole consists of a cluster of tiny, elongated, hollow gas-filled *vesicles*; each vesicle, which has a protein wall, is commonly about 70 nm in diameter. Gas vacuoles affect buoyancy in free-floating cells; for cells in a lake or river, buoyancy affects the intensity of the received light— important in the ecology of photosynthetic bacteria (Chapter 10).

2.2.6 Carboxysomes

Carboxysomes are intracellular bodies, each about 100–500 nm in diameter, which are found in many *autotrophic* bacteria, i.e. bacteria which can use carbon dioxide for most or all of their carbon requirements (Chapter 6). Each carboxysome consists of a membranous sac or shell containing many copies of an enzyme (RuBisCO) involved in the 'fixation' of atmospheric carbon dioxide.

2.2.7 Thylakoids

Thylakoids are flattened, intracellular membranous sacs which occur in most cyanobacteria; they usually occur close to, and parallel with, the cell envelope—but seem to be structurally distinct from the cytoplasmic membrane. Thylakoid membranes contain chlorophylls etc. and are the sites of photosynthesis (Chapter 5); in at least some cases they are also the sites of respiratory activity (Chapter 5).

Structures similar to thylakoids (but called *chlorosomes* or *chlorobium vesicles*) are formed by bacteria of the suborder Chlorobiineae; their functions are similar to those of thylakoids.

2.2.8 Cytoplasmic membrane

The cytoplasmic membrane (CM) is a double layer of lipid molecules, about 7–8 nm thick, in which protein molecules are partly or wholly embedded — some proteins spanning the entire thickness of the membrane (Fig. 2.4); the arrangement of lipid molecules is such that the inner and outer faces of the membrane are hydrophilic ('water-loving')—stained darkly in Plate 2 (top, left)—while the interior is hydrophobic.

The lipids are mainly phospholipids. One phospholipid, phosphatidyl-glycerol (Fig. 2.5), seems to occur in most eubacteria; the presence of other types of lipid depends mainly on whether a given bacterium belongs to one or other of two broad categories: the Gram-positive (G+ve) and Gram-negative (G−ve) bacteria—see section 2.2.9. Phosphatidylethanolamine (Fig. 2.5) is generally more common and abundant in G−ve bacteria. Phosphatidylcholine (*lecithin*) (Fig. 2.5) occurs in some G−ve bacteria but not in G+ve ones. Small amounts of glycolipids are common in bacterial CMs; sphingolipids are rare, and sterols occur in bacteria of the family Mycoplasmataceae. In *E. coli* the main lipid is phosphatidylethanolamine; phosphatidylglycerol and diphosphatidylglycerol (DPG, *cardiolipin*) are minor components.

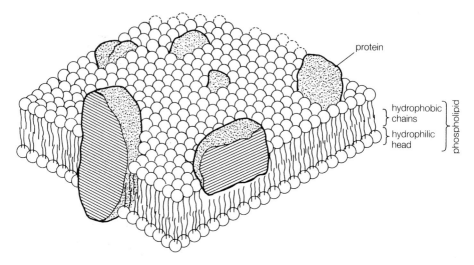

Fig. 2.4 Cytoplasmic membrane (diagrammatic): protein molecules in a fluid bilayer of phospholipid molecules—the so-called 'fluid mosaic model'. Both surfaces of the membrane are hydrophilic; the interior is hydrophobic. (Modified from Rose, *Chemical Microbiology* (3rd edition), courtesy of Butterworths Ltd.)

Archaebacterial membranes contain isoprenoid, ether-linked lipids rather than the ester-linked, fatty acid lipids described above.

The membrane is not a rigid structure: the lipid molecules are actually in a fluid state; it 'hangs together' as a result of inter-molecular forces.

The proteins include various enzymes (e.g. 'penicillin-binding proteins' involved in the synthesis of the cell envelope polymer *peptidoglycan*—sections 2.2.9.1 and 2.2.9.2); components of *transport systems* (which 'transport' ions and molecules across the membrane); and components of energy-converting systems such as *ATPases* and *electron transport chains* (Chapter 5). In at least some bacteria there are also 'sensory' proteins which detect changes in the external environment.

The CM is not freely permeable to most molecules. Some small, uncharged or hydrophobic molecules—e.g. O_2, CO_2, NH_3 (but not NH_4^+) and water—can pass through more or less freely. Other molecules (including e.g. nutrients) and ions have to be transported across by special mechanisms, some of which require expenditure of energy by the cell; these mechanisms may allow the cell to accumulate a particular substance to a concentration far greater than that which occurs in the surrounding environment.

2.2.8.1 Protoplasts

If a cell loses its cell wall (Fig. 2.2) the resulting structure is called a *protoplast*. Although bounded only by the CM, a protoplast can survive (in the laboratory) and can carry out many of the processes of a normal living cell. However, if a protoplast be suspended in a medium more dilute than its cytoplasm, water will pass in through the CM (by osmosis) and the protoplast will swell and burst—an event known as *osmotic lysis*. In an intact bacterium, the delicate protoplast is usually saved from osmotic lysis by the mechanical strength of the cell wall (section 2.2.9).

2.2.9 Cell wall

In most bacteria a tough outer layer—the cell wall (Fig. 2.2)—protects the delicate protoplast (section 2.2.8.1) from mechanical damage and osmotic lysis; it also determines a cell's shape: an *isolated* protoplast is spherical, regardless of the shape of its original cell. Additionally, the cell wall acts as a 'molecular sieve'—a permeability barrier which excludes various molecules (including some antibiotics). However, the cell wall should not be thought of merely as an 'inert box' enclosing a living cell: it also plays an active role e.g. in regulating the cell's uptake of ions and molecules.

The cell walls of different species may differ greatly in thickness, structure and composition. However, among the eubacteria there are only two major types of cell wall; whether a given cell has one or the other type

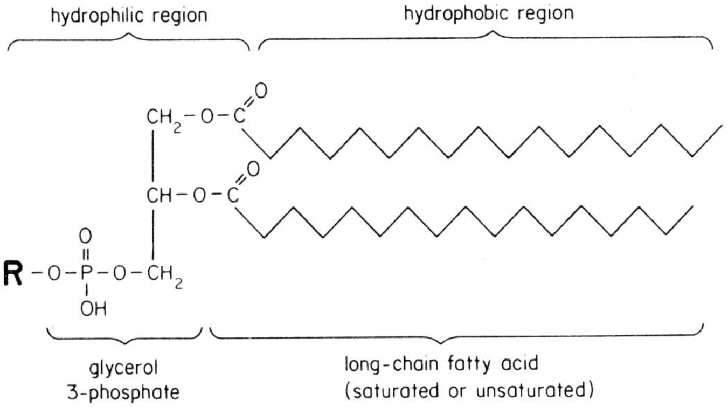

R = $^+NH_3.CH_2.CH_2-$ in phosphatidylethanolamine

R = $CH_2OH.CHOH.CH_2-$ in phosphatidylglycerol

R = $^+N(CH_3)_3.CH_2.CH_2-$ in phosphatidylcholine

Fig. 2.5 Some of the phospholipids in bacterial cytoplasmic membranes. The bond at 'R' is an ester bond. The structural formula is that of phosphatidic acid.

of wall can generally be determined by the cell's reaction to certain dyes. Thus, when stained with crystal violet and iodine, cells with one type of wall retain the dye even when treated with solvents such as acetone or ethanol; cells with the other type of wall do not retain the dye (i.e. they become decolorized) under similar conditions. This important staining procedure was discovered empirically in the 1880s by the Danish scientist Christian Gram; the *Gram stain* is described in Chapter 14. Bacteria which retain the dye (and which have one type of wall) are called *Gram-positive* bacteria; those which can be decolorized (and which have the other type of wall) are called *Gram-negative* bacteria. The cell walls of Gram-positive and Gram-negative bacteria are described below.

2.2.9.1   Gram-positive-type cell walls

The Gram-positive-type wall is relatively thick (about 30–100 nm) and it generally has a simple, uniform appearance under the electron microscope. Some 40–80% of the wall is made of a tough, complex polymer, *peptidoglycan*. Essentially, peptidoglycan consists of linear heteropolysaccharide chains that are cross-linked (by peptides) to form a three-dimensional net-like structure (the 'sacculus') which envelops the protoplast. Covalently bound to peptidoglycan are compounds such as *teichoic acids*: typically, substituted polymers of glycerol phosphate or ribitol phosphate. In some bacteria (e.g. *Mycobacterium*) the wall contains lipids, while in others (e.g. strains of *Streptococcus*) it contains carbohydrates.

The composition of the wall can vary with growth conditions; for example, in *Bacillus*, the availability of phosphate affects the amount of cell wall teichoic acids.

2.2.9.2   Gram-negative-type cell walls

The Gram-negative-type wall (20–30 nm thick) has a distinctly layered appearance under the electron microscope. The inner layer (Fig. 2.6) — nearest the cytoplasmic membrane—consists of a 'periplasmic gel' of peptidoglycan (about 1–10% of the dry weight of the cell wall). The composition of the peptidoglycan in *E. coli* is shown in Fig. 2.7.

The outer layer of the wall, the so-called *outer membrane* (Fig. 2.6, Plate 2: top, left), is essentially a protein-containing lipid bilayer—i.e. it resembles the cytoplasmic membrane. However, while the inward-facing lipids are phospholipids, the outward-facing lipids are macromolecules called *lipopolysaccharides* (LPS). Lipid A (Fig. 2.6) is a glycolipid. The core oligosaccharide (in e.g. *E. coli* and related bacteria) contains glucose and galactose residues and substituted residues of other sugars (including heptose phosphate). The O-specific chains, which form the outermost part of the cell wall, consist of linear or branched chains of oligosaccharide

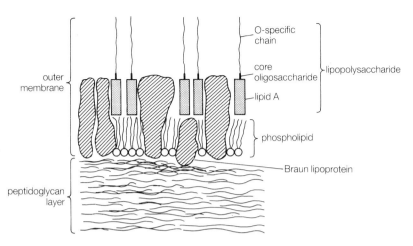

Fig. 2.6 The Gram-negative-type cell wall (diagrammatic). O-specific chains form the outermost part of the wall. The outer membrane is linked covalently to peptidoglycan through the Braun lipoproteins.

subunits; the chemical composition of the O-specific chain can vary from strain to strain, and this is exploited in the identification of particular strains by serology (section 16.1.4.2).

About half the mass of the outer membrane consists of proteins; in e.g. *E. coli* and related bacteria these include thousands of molecules of the *Braun lipoprotein* (Fig. 2.6) which are each linked covalently to the underlying peptidoglycan. There are also enzymes, proteins involved in specific uptake ('transport') mechanisms, and *porins*. A porin is one of three (sometimes two) similar protein molecules which, together, span the thickness of the membrane to form a water-filled 'pore'; such 'pores' allow certain ions and molecules to pass through the outer membrane. Porins are linked ionically to the peptidoglycan.

Components of the outer membrane are held together by ionic and other interactions. Adjacent core oligosaccharides appear to be linked by divalent cations, particularly Mg^{2+} and Ca^{2+}; experimental removal of these cations (by ion chelators) can disrupt the outer membrane. Lipid A is hydrophobically bound to the fatty acid residues of phospholipids. Some proteins appear to be linked to the core oligosaccharides.

The outer membrane is generally permeable to small ions and to small hydrophilic molecules—but much less permeable (or impermeable) to hydrophobic or amphipathic molecules.

2.2.9.3 *Archaebacterial cell walls*

In some archaebacteria (e.g. species of *Halobacterium, Methanococcus* and

Thermoproteus) the wall consists mainly or solely of a so-called 'S layer' (section 2.2.12) closely associated with the cytoplasmic membrane. In e.g. *Methanobacterium*, the wall contains *pseudomurein*: a peptidoglycan-like polymer in which the backbone chains contain N-acetyl-D-glucosamine (and/or N-acetyl-D-galactosamine, depending on species) and N-acetyl-L-talosaminuronic acid. (Unlike peptidoglycan, pseudomurein is not cleaved by the enzyme *lysozyme*.)

2.2.9.4   Layers external to the cell wall

In some bacteria there are one or more layers external to the wall; these layers include e.g. capsules, S layers and M proteins (see later).

2.2.9.5   Wall-less bacteria

Bacteria which lack cell walls are found among the eubacteria (e.g. *Mycoplasma*) and the archaebacteria (e.g. *Thermoplasma*).

Plate 1.   Some shapes, sizes and arrangements of bacteria. *Top left.* A typical view of *Helicobacter pylori* (×21 000), a helical (twisted) bacillus from the human gastrointestinal tract. This cell has three flagella (Fig. 2.2, section 2.2.14), each flagellum being covered by an extension of the cell's outer membrane (section 2.2.9.2); the bulbous structures at/near the ends of the flagella appear to be regions where the membrane has 'ballooned out'—due perhaps to the techniques used in electron microscopy. Each flagellum is about 3 μm long. *Top right.* A multicellular, filamentous bacterium, *Simonsiella*, which is present as part of the mouth microflora in about 25% of humans. This section, stained with ruthenium red, shows the filament adhering to the surface of a buccal epithelial cell. The filament is about 3.5 μm in length. *Centre.* In a 'filled' (i.e. repaired) tooth: bacteria colonizing the small space between the filling material and the wall of the cavity. Each of the 'corn-cob' structures is composed of a mass of small cocci (each less than 1 μm in diameter) clustered around the end of a hypha—perhaps indicating a symbiotic relationship between the two types of bacterium. *Bottom left.* Cells of *Aquaspirillum peregrinum*, a motile, nitrogen-fixing bacterium found in various freshwater habitats. The cell on the right is about 7.3 μm in (axial) length, 0.6 μm in thickness. *Bottom right.* A single cell of *Escherichia coli*: a straight, round-ended bacillus about 2 μm in length. This particular cell has a fragment of pilus (section 2.2.14.3) close to one side; certain viruses (Chapter 9) have been used to 'label' the pilus by forming a coating on it (arrowheads).

Photograph of *Helicobacter pylori* courtesy of Dr Alan Curry, PHLS, Withington Hospital, Manchester. *Simonsiella* courtesy of Dr Caroline Pankhurst, King's College, London, and Blackwell Scientific Publications Ltd. 'Corn-cob' bacteria courtesy of Prof. Dr Wolfgang Klimm, Medizinische Akademie 'Carl Gustav Carus', Dresden, Germany. *Aquaspirillum peregrinum* courtesy of Dr Hisanori Konishi, Yamaguchi University School of Medicine, Ube, Japan, and the Society for General Microbiology. *Escherichia coli* courtesy of Dr Markus B. Dürrenberger, University of Zürich, Switzerland.

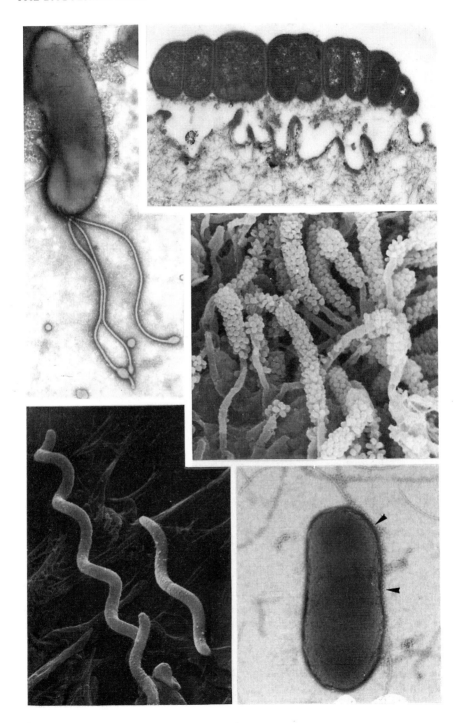

2.2.10   Adhesion sites (in Gram-negative bacteria)

Adhesion sites are localized 'fusions' between the outer membrane and the cytoplasmic membrane (Plate 2: top, right). They appear to be physiologically important regions of the cell envelope, and may serve e.g. as (i) sites where proteins (synthesized on ribosomes in the cytoplasm) are exported to (or through) the outer membrane, and (ii) anchorage sites for certain appendages called *pili* (see later).

2.2.11   Capsules and slime layers

In some bacteria the outer surface of the cell wall is covered with a layer of material called a *capsule*. A *macrocapsule*, or 'true' capsule, is thick enough to be seen (in suitable preparations) under the ordinary light microscope (i.e. thicker than about 0.2 μm), while a *microcapsule* can be detected only by electron microscopy or e.g. serological techniques (Plate 2: centre, and top left). A *slime layer* is a watery secretion which adheres loosely to the cell wall; it commonly diffuses into the medium when a cell is growing in a liquid environment.

Capsules are composed mainly of water; the organic part is usually a homopolysaccharide (e.g. cellulose, dextran) or a heteropolysaccharide (e.g. alginate, colanic acid, hyaluronic acid), but in some strains of e.g. *Bacillus anthracis* the capsule is a homopolymer of D-glutamic acid. Species of *Xanthobacter* can form an α-polyglutamine capsule together with copious polysaccharide slime. Capsule-to-wall binding may be ionic and/or covalent.

Fig. 2.7   Peptidoglycan. The structures shown are typical of those in *Escherichia coli*; similar types of peptidoglycan are found in many other Gram-negative bacteria.

(a) The (three-dimensional) net-like peptidoglycan molecule: backbone chains of alternating residues of N-acetylglucosamine (G) and N-acetylmuramic acid (M) are held together by short peptide cross-links.

(b) Part of adjacent backbone chains, showing the peptide cross-link. Here, each N-acetylmuramyl residue bears a short tetrapeptide: L-alanine–D-glutamic acid–*meso*-diaminopimelic acid–D-alanine (shown in dotted boxes); the cross-link is made covalently between the D-alanine of one tetrapeptide and the *meso*-DAP of the other. (The numbers in italic are used to refer to particular carbon atoms within the molecule.)

In Gram-positive bacteria the peptidoglycan often differs from that shown above. For example, residues of N-glycollylmuramic acid occur in the backbone chains of *Mycobacterium*, and there are many differences in the types and positions of the cross-links.

Synthesis of peptidoglycan depends on the enzymic roles of so-called penicillin-binding proteins (PBPs): proteins which occur in the cell envelope; PBPs can be inactivated by penicillins and similar antibiotics.

The enzyme *lysozyme* hydrolyses the N-acetylmuramyl-(1→4) linkages in the backbone chain; such activity can weaken the cell envelope.

(a)

(b)

Capsules have various functions. For example, they may (i) help to prevent desiccation; (ii) act as a permeability barrier to toxic metal ions; (iii) prevent infection by bacteriophages (Chapter 9); (iv) act as a nutrient reserve; (v) promote adhesion—important e.g. in those bacteria which form dental plaque; and (vi) help the cell to avoid phagocytosis. In pathogenic bacteria, capsule formation often correlates with pathogenicity (i.e. the ability to cause disease): in a given, normally capsulated pathogen, capsule-less strains are typically non-pathogenic.

Some secreted polysaccharides are used industrially. For example, the heteropolysaccharide *xanthan gum* (produced by strains of *Xanthomonas campestris*) is used in the food industry as a gelling agent, gel stabilizer, thickener, and inhibitor of crystallization.

2.2.12   S layers

Some cells have so-called 'S layers'—usually as the outermost layer of the cell; an S layer consists of a repeating pattern of protein or glycoprotein subunits arranged e.g. in squares or hexagons. In those eubacteria which have an S layer, the S layer usually overlays the cell wall, e.g., in Gram-negative bacteria it covers the outer membrane. In some archaebacteria the S layer *is* the cell wall, i.e. it overlays the cytoplasmic membrane.

Double S layers occur e.g. in strains of *Aquaspirillum* and *Bacillus*. In one strain an S layer has been reported to occur on *both* sides of the cell wall [Wahlberg *et al.* (1987) FEMS 40 75–79].

2.2.13   M proteins

'M proteins' are proteins which form a thin layer on the cell wall in certain virulent strains of *Streptococcus*; this layer makes the cells less susceptible to phagocytosis, thus contributing to the organism's virulence.

2.2.14   Flagella, fimbriae and pili

In many bacteria there are fine, hair-like proteinaceous filaments extending from the cell surface; these filaments can be divided into three main types: flagella, fimbriae and pili.

2.2.14.1   Flagella

Flagella (singular: *flagellum*) enable a cell to swim through a liquid medium, i.e. they are involved in cell motility (section 2.2.15). Depending on species, a cell may have a single flagellum (*monotrichous* arrangement); one flagellum at each end (*amphitrichous* arrangement); a tuft of flagella at one or both ends (monopolar or bipolar *lophotrichous* arrangement); or flagella which arise all over the cell surface (*peritrichous* arrangement) (see Fig. 2.1).

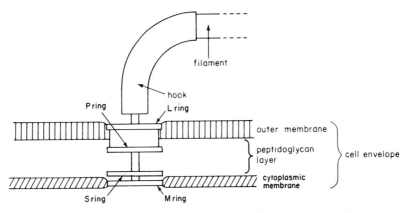

Fig. 2.8 The attachment of a flagellum to the cell envelope of a Gram-negative bacterium (*Escherichia coli*). The four parallel rings (L, P, S and M) and the core form the basal body (section 2.2.14.1).

Each flagellum consists of a *filament*, a *hook* and a *basal body* (Fig. 2.2). The filament is helical, $5-20 \times 0.02 \, \mu m$, and is composed of eleven protein fibrils arranged like the strands of a rope; a fine channel runs through the axis of the filament. (In some species of e.g. *Pseudomonas* and *Vibrio*, and in *Helicobacter pylori*, the filament is sheathed, i.e. covered by an extension of the outer membrane—see e.g. Plate 1: top, left). The hook is a curved proteinaceous structure (Fig. 2.8). The basal body consists of several co-axial ring-shaped components on a central, hollow rod-shaped core (Fig. 2.8); precise details vary according to species: in *Escherichia coli* and related bacteria there are four rings, in *Bacillus subtilis* two rings, and in *Caulobacter crescentus* five rings. In *E. coli* (according to one model) the L, P and S rings are fixed—the central 'shaft' (together with hook and filament) *rotating* as the M ring rotates. Flagellar rotation has been demonstrated by anchoring the free end of a flagellum to a glass surface and observing rotation of the cell body. Rotation needs energy in the form of an ion gradient across the cytoplasmic membrane (Chapter 5).

Flagella can be seen by light microscopy only after suitable staining.

In spirochaetes, so-called *periplasmic flagella* (similar to those described) occur *between* the layers of the cell envelope, one or more of these flagella arising from each end of the cell; the number of periplasmic flagella per cell varies with genus and is a stable characteristic.

2.2.14.2 Fimbriae

Fimbriae (singular: *fimbria*) are very thin, proteinaceous filaments (typically $0.1-5 \, \mu m \times 2-10 \, nm$) which occur mainly (and quite commonly) on Gram-negative bacteria. They may occur all over the cell surface (Plate 2: bottom, right) or may be localized. Each fimbria consists of linear repeating protein

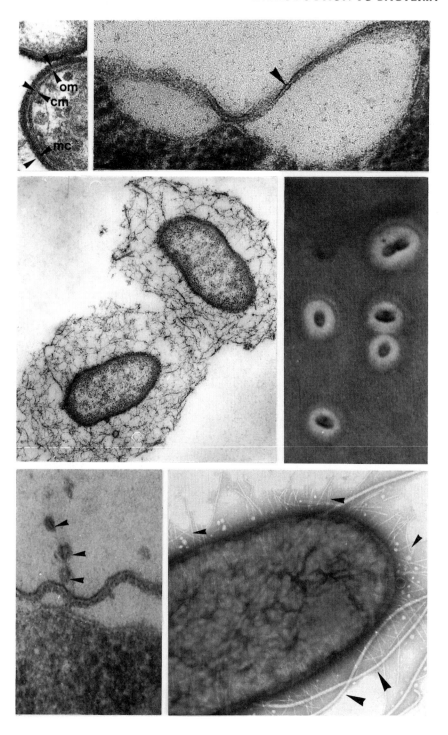

subunits; often the subunits are rich in non-polar amino acids, so that fimbriate cells (those having fimbriae) tend to have surfaces which are more hydrophobic than those of afimbriate cells. In Gram-negative bacteria at least some types of fimbria seem to be attached only to the outer membrane (compare flagellar attachment).

In Gram-negative bacteria many types of fimbria promote cell-to-cell adhesion or cell-to-surface adhesion; in pathogenic species (e.g. *Neisseria gonorrhoeae*) this can be important—enabling the bacteria to stick to certain tissues. Bacilli with polar (terminal) fimbriae exhibit so-called 'twitching motility' (section 2.2.15.1).

2.2.14.3 *Pili*

Pili (singular: *pilus*) are elongated or hair-like proteinaceous structures which project from the cell surface; they are found specifically on those Gram-negative cells which have the ability to transfer DNA to other cells by *conjugation* (Chapter 8)—a process in which (in at least some cases) the pili themselves play an essential role. The genes encoding pili occur in genetic

---

Plate 2. The cell envelope and surface structures in some Gram-negative bacteria. *Top left.* The outer membrane (om) and cytoplasmic membrane (cm), and a microcapsule (mc), in a thin section of *Bacteroides fragilis* (x 52 000). The microcapsule seen in this electronmicrograph would not be seen by light microscopy. *Top right.* The cell envelope in a thin section of a plasmolysed cell of *Escherichia coli* (x 100 000). (Plasmolysis, in which water is withdrawn from the cell, causes the cytoplasmic membrane to shrink away from the outer membrane.) Here, gaps are seen between the outer membrane (arrowhead) and the cytoplasmic membrane below it. Near the centre of the photograph, the outer membrane is joined to the cytoplasmic membrane; this 'fused' region is called an adhesion site. *Centre left.* Two capsulated cells of *B. fragilis* as seen by electron microscopy (x 28 000). The macrocapsules have been stained with ruthenium red. *Centre right.* Light microscopy of cells of *B. fragilis* from the same culture as those seen to the left (approx. x 1000). Here, the background has been stained darkly with eosin–carbolfuchsin ('negative staining') so that each capsule can be seen as a bright 'halo' surrounding its cell. The cell at the top is dividing. *Bottom left.* An F pilus (a specialized, hair-like appendage) arising from the cell envelope in *E. coli* (x 150 000). The F pilus itself (less than 10 nm in diameter) is barely visible; however, in order to make it detectable, it has been 'labelled' with a particular bacteriophage (MS2) which binds specifically to the sides of these pili. Here, three MS2 bacteriophages (each about 25 nm in diameter) have bound, close together, along part of the length of the F pilus; each bacteriophage is indicated by an arrowhead. *Bottom right.* Part of an *E. coli* cell specially stained to show the large number of fimbriae (indicated by small arrowheads) (x 54 000). The larger arrowheads point to fragments of flagella.

Photographs of *B. fragilis* courtesy of Dr Sheila Patrick, Queen's University of Belfast. Adhesion site and F pilus in *E. coli* courtesy of Dr Manfred E. Bayer, Fox Chase Cancer Center, Philadelphia. *E. coli* fimbriae courtesy of Dr Anne Field, Public Health Laboratory Service, London.

elements called *plasmids* (Chapter 7). Commonly, only one or a few pili occur on a given cell.

The various types of pili differ e.g. in size and shape: for example, some are long, thin and flexible, while others are short, rigid and nail-like; the type of pilus correlates with the physical conditions under which conjugation can take place (Chapter 8). The best-studied flexible pilus, the F *pilus*, is one to several micrometres long, 8–9 nm in diameter; a 'labelled' F pilus is seen in Plate 2 (bottom, left). The F pilus, a tubular structure with an axial canal about 2.5 nm in diameter, is made of helically arranged protein subunits.

2.2.15   Motility and chemotaxis

Many bacteria are *motile*, i.e. they can actively move about in liquid media. Not only can they move, they can also move towards better sources of nutrients and away from harmful substances; such a directional response is called *chemotaxis*.

2.2.15.1   *Motility*

In most cases motility is due to the possession of one or more flagella (section 2.2.14.1). Flagellar motility involves energy-requiring *rotation* of the flagellum from the basal body.

Peritrichously flagellated cells (e.g. *E. coli*) can reach speeds of about 30 μm/s: the flagella bunch together at one end of the cell, all rotating in a counterclockwise direction, and the cell moves with the opposite end leading. Under uniform conditions, such 'smooth swimming' is interrupted about once per second by *tumbling*: a brief random movement caused by clockwise rotation of the flagella; tumbling lasts about 0.1 s. Alternate swimming and tumbling results in a three-dimensional 'random walk': the cell moves in a series of straight lines in randomly determined directions.

Monotrichously flagellated cells (e.g. strains of *Pseudomonas aeruginosa*) can reach speeds of 60–70 μm/s. In these cells the flagellum rotates clockwise and counterclockwise for roughly equal periods of time; during change-over, conformational changes in the filament and/or hook of the flagellum may cause randomization of direction—similar to that caused by tumbling in peritrichous cells.

The spirochaete's screw-like motion is thought to involve rotation of the periplasmic flagella (causing wave-like ripples in the cell envelope); spirochaetes can reverse direction and 'flex'—the latter movement probably permitting randomization of direction.

So-called *gliding motility* occurs in certain bacteria which lack flagella—e.g. species of *Beggiatoa* and *Oscillatoria*. It is a smooth form of motion which appears to occur only when cells are in contact with a solid surface; a gliding cell leaves a 'slime trail' but the mechanism of gliding is not known.

Cells with polar (localized) fimbriae (section 2.2.14.2) exhibit *twitching motility* in thin films of water. The mechanism is not known, but it seems to be a passive form of movement (not true motility) involving interaction between the cell and extracellular physicochemical forces. Twitching occurs e.g. in species of *Acinetobacter, Neisseria* and *Pseudomonas*—i.e. in both flagellate and non-flagellate cells.

2.2.15.2   Chemotaxis

In a chemically uniform environment, flagellated cells typically adopt a 'random walk' (section 2.2.15.1). Suppose, however, that the concentration of nutrients increases in a certain direction; can a cell—which changes direction *randomly*—travel towards the higher concentration of nutrients? It can, simply by tumbling less frequently when moving in the 'right' direction, but at normal frequency when moving in other directions. In this way, more time is spent going in the 'right' direction—so that the cell's *overall* (net) movement will be towards the higher concentration of nutrients.

How does the cell 'sense' different concentrations, and how does it control its rate of tumbling? Essentially—according to one model—special 'sensor' proteins (in the cell envelope) respond to concentration gradients by controlling the numbers of smaller, 'effector' molecules in the cytoplasm. Depending on their concentration, the effector molecules (in *E. coli*) can either (i) favour counterclockwise flagellar rotation (and, hence, less tumbling) or (ii) clockwise rotation (and, hence, more tumbling). Attractants (e.g. nutrients) and repellents generally· affect sensors in opposite ways.

2.3   TRICHOMES AND COENOCYTIC BACTERIA

Earlier it was mentioned that most bacteria can live as single, autonomous cells. Thus, for example, each cell in a chain of streptococci leads an essentially independent life—except, of course, in that each cell shares a micro-environment with its neighbour(s). Some bacteria, however, normally exist in *trichomes* or as *coenocytic* organisms.

2.3.1   Trichomes

A trichome is a row of cells which have remained attached to one another following successive cell divisions; the cells are separated by *septa* (cross-walls, singular: *septum*), but, in at least some trichomes, adjacent cells communicate with one another via small pores (*microplasmodesmata*). (In a simple 'chain' of cells—as formed e.g. by some streptococci— such pores are not formed.) The positions of the septa may or may not be obvious (as constrictions) from the outside of the trichome. The cells of a trichome may or may not be covered by a common sheath. Trichomes are formed by many cyanobacteria and e.g. by species of *Beggiatoa*.

2.3.2   Coenocytic bacteria

The filamentous actinomycete *Streptomyces*, and some other bacteria, form tube-like hyphae which lack septa, the cytoplasm being continuous from one nucleoid to the next. Such a multinucleate organism is called a *coenocyte*.

3 Growth and reproduction

Growth in a bacterial cell involves a *co-ordinated* increase in the mass of its constituent parts; it is not simply an increase in total mass since this could be due, for example, to the accumulation of a storage compound within the cell.

Usually, growth leads to the division of a cell into two similar or identical cells. Thus, growth and reproduction are closely linked in bacteria, and the term 'growth' is generally used to cover both processes.

3.1 CONDITIONS FOR GROWTH

Bacteria grow only if their environment is suitable; if it's not optimal, growth may occur at a lower rate or not at all—or the bacteria may die, depending on species and conditions.

Essential requirements for growth include (i) a supply of suitable nutrients; (ii) a source of energy; (iii) water; (iv) an appropriate temperature; (v) an appropriate pH; (vi) appropriate levels (or the absence of) oxygen. Of course, none of these factors operates in isolation: an alteration in one may enhance or reduce the effects of another; for example, the highest temperature at which a bacterium can grow may well be lowered if the pH of the environment is made non-optimal.

3.1.1 Nutrients

Cells need nutrients as raw materials for growth, maintenance and division. As a group, the bacteria use an enormous range of compounds as nutrients; these include various sugars and other carbohydrates, amino acids, sterols, alcohols, hydrocarbons, methane, inorganic salts and carbon dioxide. However, no individual bacterium can use all of these compounds: it hasn't the range of enzymes to deal with them all, and (in any case) its cell envelope does not have uptake ('transport') systems for all of them. A given type of bacterium typically uses only a relatively limited range of compounds.

Whatever the organism, cells need sources of carbon, nitrogen, phosphorus, sulphur and other materials from which living matter is made. Some bacteria obtain all nutritional requirements from simple inorganic salts and substances such as carbon dioxide and ammonia; others need—to varying extents—more or less complex organic compounds derived from other organisms. Some important aspects of nutrition are discussed in Chapters 6 and 10.

3.1.2   Energy

Energy is needed for many of the essential chemical reactions which go on in a living cell; it is also needed e.g. for flagellar motility and for the uptake of various nutrients. All of this energy is obtained from sources in the environment. In photosynthetic species, energy is derived mainly or solely from light, while chemotrophic species obtain energy by 'processing' chemicals taken from the environment. Some species can use both methods. Energy is discussed in Chapter 5.

3.1.3   Water

Some 80% or more of the mass of a bacterium is water, and, during growth, nutrients and waste products enter and leave the cell, respectively, *in solution;* hence, bacteria can grow only in or on materials which have adequate free (available) water. (Not all the water in a given material is necessarily available for bacterial growth; some, for example, may be 'tied up' by hydrophilic gels or by ions in solution.)

3.1.4   Temperature

Generally, for a given type of bacterium, growth proceeds most rapidly at a particular temperature: the *optimum growth temperature;* the rate of growth tails off as temperatures increase or decrease from the optimum. For any given bacterium there are maximum and minimum temperatures beyond which growth will not occur.

Thermophilic bacteria are those whose optimum growth temperature is >45 °C. These *thermophiles* occur e.g. in composts, hot springs and hydrothermal vents on the ocean floor; they include species of *Thermobacteroides* (opt. 55–70 °C), *Thermomicrobium* (opt. 70–75 °C) and *Pyrodictium* (opt. 105 °C).

Thermoduric bacteria can survive—though not necessarily grow—at temperatures which would normally kill most other vegetative (i.e. growing) bacteria. In dairy bacteriology, 'thermoduric' bacteria are those which survive pasteurization (section 12.2.1.1).

Mesophilic bacteria grow optimally at temperatures between 15 and 45 °C. The *mesophiles* live in a wide range of habitats, and they include those bacteria which are pathogens in man and other animals.

Psychrophilic bacteria grow optimally at or below 15 °C, do not grow above about 20 °C, and have a lower limit for growth of 0 °C or below. The *psychrophiles* occur e.g. in polar seas.

Psychrotrophic bacteria can grow at low temperatures (e.g. 0–5 °C), but they grow optimally above 15 °C, with an upper limit for growth >20 °C.

3.1.5   pH

Most bacteria grow best at or near pH 7 (neutral), and the majority cannot grow under strongly acidic or strongly alkaline conditions. However, some (found e.g. in mine drainage and in certain hot springs) not only tolerate but actually 'prefer' acidic or highly acidic conditions; these *acidophiles* include *Thiobacillus thiooxidans* (opt. pH 2–4), species of *Sulfolobus* (which grow at pH 1–5) and *Thermoplasma acidophilum* (pH 0.8–3).

Alkalophiles grow optimally under alkaline conditions—typically above pH 8. *Thermomicrobium roseum* (opt. pH 8.2–8.5) occurs e.g. in hot springs, and *Exiguobacterium aurantiacum* (opt. pH 8.5 and 9.5) has been found in potato-processing effluents; other alkalophiles occur in natural alkaline lakes.

Acidophiles and alkalophiles may grow slowly—or not at all—at pH 7.

3.1.6   Oxygen

Some bacteria need oxygen for growth. Others need the *absence* of oxygen for growth. Yet others can grow regardless of the presence or absence of oxygen.

Bacteria which *must* have oxygen for growth are called 'strict' or 'obligate' *aerobes* to emphasize their absolute need for oxygen.

Strict or obligate *anaerobes* grow only when oxygen is absent; these organisms occur e.g. in river mud and in the rumen.

Bacteria which normally grow in the presence of oxygen but which can still grow under anaerobic conditions (i.e. in the absence of oxygen) are called *facultative anaerobes*. Similarly, those which normally grow anaerobically but which can grow in the presence of oxygen are called *facultative aerobes*.

Microaerophilic bacteria generally grow best when the concentration of oxygen is (usually much) lower than it is in air.

3.1.7   Inorganic ions

All bacteria need certain inorganic ions (e.g. those of chlorine and magnesium) in low concentrations, higher concentrations usually inhibiting growth. The ions have various functions—e.g. magnesium in the outer membrane (section 2.2.9.2), iron in cytochromes (section 5.1.1.2) and in a range of enzymes, and manganese and nickel in enzymes or enzyme systems.

Some bacteria (the *halophiles*) grow only in the presence of a high concentration of electrolyte (usually NaCl). Archaebacteria of the family Halobacteriaceae, which occur in salt lakes and salted fish etc., are examples of extreme halophiles: they need at least 1.5 M NaCl for growth, 3–4 M NaCl for good growth. The electrolyte serves to maintain the structure of e.g. ribosomes and the cell envelope; in dilute solutions the cells of some species break open due to weakening of the cell envelope.

Halotolerant bacteria are non-halophiles which can grow in electrolyte up to about 2.5 M; they include many strains of *Staphylococcus*.

3.2 GROWTH IN A SINGLE CELL

In a growing cell there is a co-ordinated increase in the mass of the component parts. 'Co-ordinated' does not mean that all the parts are made simultaneously: some are synthesized more or less continually, but others are made in a definite sequence during certain fixed periods. The cycle of events in which a cell grows, and divides into two daughter cells, is called the *cell cycle*.

3.2.1   Cell cycle

Most studies on the bacterial cell cycle have been carried out on *Escherichia coli* (a Gram-negative bacillus). During growth, the cells of *E. coli* increase in size mainly by elongation, and this clearly requires the synthesis of new wall and membrane materials; during continuous, steady growth, such synthesis is presumably continuous throughout the cell cycle. A model for the synthesis of the cell surface during the cell cycle has recently been published [Cooper (1991) MR 55 649–674]. In this model, peptidoglycan is incorporated diffusely in the lateral wall, and is synthesized preferentially at the poles of the cell. The cytoplasmic membrane may develop in a manner similar to that of peptidoglycan.

Events which occur only during fixed periods in the cell cycle include replication of the chromosome(s) (Chapter 7) and, subsequently, the formation of a cross-wall (*septum*) which divides the cell into two cells; septum formation involves the ingrowth of peptidoglycan from the cell wall at a site half-way along the length of the cell. During cell division, both daughter cells receive the same number of chromosomes; this is probably due to the fact that chromosomes are associated with the cytoplasmic membrane; during membrane growth, the chromosomes may be drawn apart mechanically.

Following cell division in *E. coli*, daughter cells usually separate immediately. In other species, separation may not occur immediately—leading to one or other of the groupings mentioned in section 2.1.3.

3.2.1.1   *Helmstetter–Cooper model*

During the cell cycle it's clearly essential that chromosome replication and cell division be properly coordinated. Chromosome replication within the cell cycle is described by the Helmstetter–Cooper model. Figure 3.1 shows growth at two different rates. During slow growth, one complete

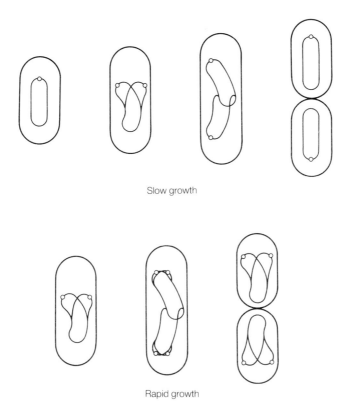

Slow growth

Rapid growth

Fig. 3.1 Growth at different rates. Cells and chromosomes are represented diagrammatically. Chromosome replication begins at the 'origin' (section 7.3)— shown as a small circle. See section 3.2.1.1 for explanation.

duplication of the chromosome coincides with one division of the cell—so that each new daughter cell receives only one chromosome. During faster growth, a new round of chromosome replication is started before the previous round has been completed (Fig. 3.1, bottom row, centre); each new daughter cell therefore receives more than one chromosome—about 1½ chromosomes in Fig. 3.1. This helps to explain why it is that the rate of growth can affect the number of chromosomes in a cell (section 2.2.1).

3.2.1.2   Morphogenes

Genetic control (Chapter 7) of structural events in the cell cycle is exercised through the *morphogenes*. In *E. coli* these genes include *envA* (involved in processing the septum), *minB* (ensuring septation in the correct part of the cell) and genes encoding the so-called penicillin-binding proteins (involved in cell envelope synthesis).

3.2.1.3 Control of the cell cycle

In the cell cycle, the occurrence of any given event could be directly dependent on the occurrence of the previous event. Alternatively, events could be controlled independently, and co-ordinated. These alternative ways of viewing the cell cycle have recently been evaluated [Nordström, Bernander & Dasgupta (1991) MM 5 769–774].

3.2.2 Modes of cell division

The division of one cell into two (typically similar or identical) cells by the formation of a septum (as in E. coli) is called binary fission; it is the commonest form of cell division in bacteria. 'Asymmetrical' binary fission, in which daughter cells are not similar to the parent cell, occurs e.g. in Caulobacter (Chapter 4).

Multiple fission involves repeated binary fission of cells within a common bag-like structure; it occurs e.g. in certain cyanobacteria.

In ternary fission, three cells are formed from one; it occurs e.g. in Pelodictyon (which forms three-dimensional networks of cells).

Budding is a form of cell division in which a daughter cell develops from the mother (parent) cell as a localized outgrowth (bud); it occurs e.g. in species of Blastobacter, Hyphomicrobium and Nitrobacter.

3.2.3 Doubling time

The time taken for one complete cell cycle (the doubling time) varies with species and with growth conditions. For a minimum doubling time, optimum growth conditions are necessary. In E. coli the minimum doubling time is about 20 minutes; in some species of Mycobacterium (for example) it is many hours.

3.3 GROWTH IN BACTERIAL POPULATIONS

Since, following cell division, each daughter cell can itself grow and divide, one cell can quickly give rise to a large population of cells if conditions are favourable. Given suitable conditions, such populations may develop either on solid surfaces or within the body of a liquid. In bacteriology, any solid or liquid specially prepared for bacterial growth is called a medium (section 14.2).

3.3.1 Growth on a solid medium

One common type of 'solid' medium, widely used in bacteriological laboratories, is a jelly-like substance (an agar gel) containing nutrients and

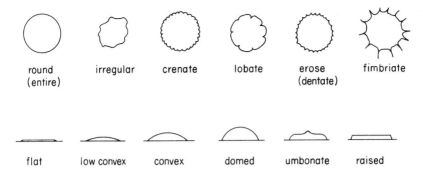

Fig. 3.2 Shapes of bacterial colonies. The upper row shows the outline or 'edge' of some colonies as seen from above. The lower row shows the 'elevation' of some colonies as seen from one side. A colony with, say, a round (entire) outline may have any one of the elevations, depending e.g. on the species of bacterium.

other ingredients. Suppose that a single bacterial cell is placed on the surface of such a medium and given everything necessary for growth and division. The cell grows, divides into two cells, and each daughter cell does the same. If growth and division continue, the progeny of the original cell eventually reach such immense numbers that they form a compact heap of cells that is usually visible to the naked eye; this mass of cells is called a *colony*.

Typically, under given conditions, each species forms colonies of characteristic size, shape (Fig. 3.2), colour and consistency; different types of colony may be formed when growth occurs on different media or when other factors differ. The size of a colony may be limited e.g. by local exhaustion of nutrients (due to the colony's own growth); for this reason, crowded colonies are generally smaller than well-spaced ones. The rate at which a colony increases in size depends on temperature and other factors. Bacteria which produce pigments generally form brightly coloured colonies (e.g. red, yellow, violet) while colonies of non-pigmented bacteria usually look grey, whitish or cream-coloured. In consistency, a colony may be mucoid (viscous, mucus-like), butyrous (butter-like), friable (crumbly) etc. and its surface may be smooth or rough, glossy or dull etc.

Instead of starting with a single cell, suppose that a very large number of cells is spread over the surface of the medium. In this case there may not be enough space for individual colonies to develop; accordingly, the progeny of all these cells will form a continuous layer of bacteria which covers the entire surface of the medium. Such continuous growth is called *confluent growth*. Confluent growth can also result when one or a few cells of a motile bacterium are deposited on the medium; following growth, the numerous progeny of these cells may swim through the surface film of moisture and eventually cover the whole surface of the medium.

3.3.2   Growth in a liquid medium

Bacteria can move freely through a liquid medium either by diffusion or, in motile species, by active locomotion; thus, as cells grow and divide, the progeny are commonly dispersed throughout the medium. Usually, as the concentration of cells increases, the medium becomes increasingly turbid (cloudy). Certain bacteria are exceptional in that they tend to form a layer (a *pellicle*) at the surface of the medium; below the pellicle the medium may be almost free of cells. Some pellicles include bacterial products as well as the bacteria themselves; for example, a tough cellulose-containing pellicle is formed by strains of *Acetobacter xylinum*.

3.3.2.1   *Batch culture*

Suppose that a few bacterial cells are introduced into a suitable liquid medium which is then held at the optimum growth temperature for that species. At regular intervals a small volume of the medium can be withdrawn and a count made of the cells it contains (counting methods: section 14.8). In this way we can follow the development of a population, i.e. the increase in cell numbers with time. By plotting the number of cells against time we obtain a *growth curve* which, for a given species growing under given conditions, has a characteristic shape.

By growing bacteria in or on a medium we produce a *culture*; thus, a culture is a liquid or solid medium containing bacteria which have grown (or are still growing) in or on that medium. The process of maintaining a particular temperature (and/or other desired conditions) for bacterial growth is called *incubation*. The initial process of adding the cells to the medium is called *inoculation*.

When bacteria are introduced into a fresh liquid medium, cell division may not begin immediately: there may be an initial *lag phase* in which little or no division occurs. During lag phase the cells are adapting to their new environment—for example, by making enzymes to utilize the newly available nutrients. The length of the lag phase will depend largely on the conditions under which the cells existed *before* they were introduced into the medium. A long lag phase will often occur if the cells had previously existed under harsh conditions, or had been growing with different nutrients or at a different temperature; the lag phase will be short (or even absent) if the cells had been growing in a similar or identical medium at the same temperature.

During the (adaptive) lag phase, molecules are being synthesized, but the increase in total mass of the cell population is not matched by an increase in cell numbers; the cells are said to be undergoing *unbalanced growth*.

Once adapted to the new medium, the cells begin to grow and divide at a rate which is maximum for the species under the existing conditions; this is the *logarithmic* (= *log*) *phase* or *exponential phase* of growth. In this phase, cell

Table 3.1 Increase in cell numbers with time for *Escherichia coli* growing under optimal conditions in the logarithmic phase

Time (minutes)	0	20	40	60	80	100	200	300
Number of generations (i.e., rounds of cell division)	0	1	2	3	4	5	10	15
Number of cells	1	2	4	8	16	32	1024	32768
Number of cells as a power of 2	2^0	2^1	2^2	2^3	2^4	2^5....	2^{10}....	2^{15}

numbers double at a constant rate (Table 3.1)—as does the mass of the population; this indicates *balanced growth*.

In the log phase of growth, a plot of cell numbers versus time gives a sharply rising curve on a simple arithmetical scale (Fig. 3.3a); clearly, such a scale would not be adequate for large numbers of cells. Is there a better way of plotting growth in the log phase? Table 3.1 (bottom row) shows that cell numbers can be expressed as powers of 2; for example, the 8 cells at 60 minutes can be written as 2^3 cells (in which 3 is the *index*). Each of the indices in Table 3.1 is, of course, the *logarithm* (to base 2, i.e. $\log_2$) of the corresponding number of cells. Now, if—instead of plotting cell numbers directly—we plot the $\log_2$ of each number, the result is a straight-line graph (Fig. 3.3b). In such a graph each unit on the $\log_2$ scale represents a doubling in cell numbers; the *doubling time* (here, the time, in minutes, needed for a doubling of cell numbers) can therefore be read off directly from the time-scale of the graph. The doubling time is also called the *generation time*.

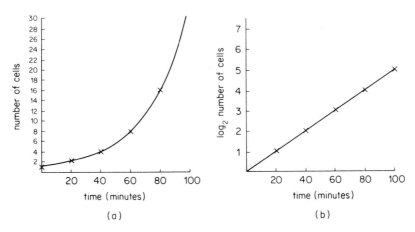

Fig. 3.3 The logarithmic (exponential) phase of growth in *Escherichia coli*. Cell numbers are plotted on (a) a simple arithmetical scale, and (b) a logarithmic ($\log_2$) scale. (Compare with Table 3.1.)

Usually it's more convenient to use $\log_{10}$ rather than $\log_2$ when constructing a growth curve. The $\log_{10}$ and $\log_2$ of any number can be interconverted by using the formula: $\log_{10} N = 0.301 \times \log_2 N$; the graph will still be a straight line—only the *slope* of the graph changes when the base is changed.

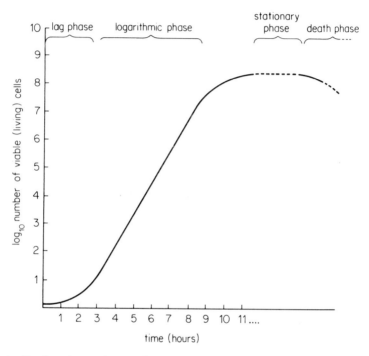

Fig. 3.4 Batch culture. A growth curve constructed for a strain of *Escherichia coli* growing in nutrient broth at 37°C.

As they grow, the cells use nutrients and they also produce waste products which accumulate in the medium. Eventually, therefore, growth slows down and stops due either to a lack of nutrients or to the accumulation of waste products (or both); the phase in which there is no overall increase in the number of living cells is called the *stationary phase*. (A recent minireview entitled 'Life after log' is worth reading [Siegele & Kolter (1992) JB *174* 345–348].) The stationary phase leads eventually to the *death phase* in which the number of living cells in the population progressively decreases.

Figure 3.4 shows the phases of growth during batch culture. 'Batch culture' is so-called because growth from lag phase to death phase occurs in the same batch of medium.

3.3.2.2 Cells in different phases of growth

Cells can undergo marked changes in their general biology and metabolism during the different phases of growth. For example, cells which—when growing—are normally killed by penicillin (section 15.4.1) are resistant to this antibiotic when growth ceases.

During periods of non-growth or growth-limitation, by-products of primary (growth-directed) metabolism may be used by the cell to synthesize so-called 'secondary metabolites' ('idiolites') which are not used for growth. In some species, idiolites include important antibiotics or toxins. [Function of secondary metabolites: Vining (1990) ARM 44 395–427.]

The production of substances during a particular phase of growth can be optimized by adding nutrient(s) at appropriate times; such *fed batch culture* is used in some industrial processes.

3.3.2.3 Continuous culture

When bacteria are grown in a fixed volume of liquid medium (as in batch culture) the composition of the medium continually changes as nutrients are used up and waste products accumulate. Batch culture is suitable for many types of study, but sometimes it is preferable that cells be grown under constant, controlled, defined conditions. This is achieved by using *continuous culture* (*continuous-flow culture, open culture*). In this process, bacteria are grown in a liquid medium within an apparatus called a *chemostat*; during growth, there is a continual inflow of fresh, sterile medium, and a simultaneous outflow —at the same rate—of culture (i.e. medium + cells). Constant and thorough agitation of the medium is necessary to ensure rapid mixing of the inflowing fresh medium with culture in the chemostat. Under these conditions, cells can exhibit continual logarithmic growth, i.e. balanced growth, for an extended period of time. Because growth is occurring under constant and defined conditions, this form of culture is useful e.g. for studies on bacterial metabolism.

In the chemostat, growth is normally kept at a *sub*maximal rate because instability tends to occur in the system at or near the maximum growth rate. The growth rate is controlled by controlling the concentration of an essential nutrient in the inflowing medium.

Under 'steady-state' conditions in a chemostat, the increase in cell numbers through growth is exactly balanced by the loss of cells from the chemostat; the mass of cells (the *biomass*) in the chemostat therefore remains constant. To achieve a steady state, the dilution rate is made equal, numerically, to the specific growth rate. The *dilution rate* (D) is given by F/V, where F is the rate at which the medium enters the chemostat $(l\,h^{-1})$ and V is the volume of the culture. The *specific growth rate* (μ) is the number of grams of biomass formed, per gram of biomass, per hour. The specific growth rate and the concentration of the growth-limiting nutrient often have the relationship predicted by the Monod equation:

$$\mu = \mu_{max} \; \frac{s}{k_s + s}$$

in which μ_{max} is the maximum growth rate, s is the concentration of the growth-limiting nutrient, and k_s is the concentration of the growth-limiting nutrient when $\mu = 0.5\,\mu_{max}$.

Clearly, D should not exceed the *critical dilution rate*, D_c, which corresponds to μ_{max}. If it does, the culture becomes progressively diluted to extinction and is said to have undergone 'wash-out'.

In practice, ideal (predictable) operation of a chemostat is not always achieved—due e.g. to less than perfect (instantaneous) mixing. Another type of problem involves the emergence of *mutants* (Chapter 8) during the long periods of growth.

3.3.2.4   *Synchronous growth*

In a population of growing bacteria, all the cells do not divide at the same instant. However, in the laboratory, we can obtain a population in which all the cells divide at approximately the same time; in such *synchronous growth* the logarithmic portion of the growth curve (Fig. 3.4) appears as a series of steps, each step representing an abrupt doubling of cell numbers.

3.4   DIAUXIC GROWTH

If a bacterium is given a mixture of two different nutrients, it may use one in preference to the other—utilization of the second beginning only after the first has been exhausted. For example, given a mixture of glucose and lactose, *Escherichia coli* will use the glucose first, starting on the lactose only when all the glucose has been used. During the transition from one nutrient to the other, growth may slow down or even stop; this pattern of growth is called *diauxie* (or diauxy). The mechanism of diauxie is discussed in section 7.8.2.1.

3.5   MEASURING GROWTH

Growth (change in cell numbers, or biomass, with time) can be measured e.g. by (i) counting cells (section 14.8); (ii) determining the increase in dry weight of biomass formed in a given time interval; (iii) monitoring the uptake (or release) of a particular substance; (iv) measuring the amount of a

radioactive substance incorporated in biomass in a given time, and
(v) *nephelometry*: measuring the increase in scattered light from a beam
passing through a liquid culture (light scattering increases as cell numbers
increase).

4 Differentiation

In most species of bacteria, major changes in form or function do not occur: progeny cells are more or less identical, both in appearance and behaviour, to their parental cells. In some bacteria, however, one type of cell can give rise to a markedly different type of cell; the timing of such *differentiation* is commonly related to conditions in the cell's environment. In the next few pages we look at some diverse examples of bacterial differentiation.

4.1 THE LIFE-CYCLE OF *Caulobacter*

Caulobacter is a Gram-negative, strictly aerobic bacterium found in soil and water. It forms two distinctly different types of cell, and the change from one type to the other is an essential part of the life-cycle (Fig. 4.1). Having a swarm cell in the life-cycle is advantageous since it allows the organism to spread to different locations. The motile daughter cell must lose its flagellum and develop a stalk (called a *prostheca*) before it can divide, and may thus be regarded as immature; the stalked (mature) mother cell can produce swarm cells but it cannot itself become a swarm cell.

A similar type of differentiation, from non-motile to motile cells, and vice versa, occurs in species of *Hyphomicrobium* and *Rhodomicrobium*.

4.2 SWARMING IN *Proteus*

Proteus is a Gram-negative bacillus found e.g. in the intestines of man and other animals. If cells of *P. mirabilis* (or of another species, *P. vulgaris*) are incubated on a suitable solid medium, the first progeny cells are short, sparsely flagellated bacilli about 2–4 μm in length; these cells form a colony in the usual way. However, after several hours of growth, some of the cells around the edge of the colony grow to lengths of 20–80 μm and develop numerous additional flagella; these cells are called *swarm cells*. The swarm cells swim out to positions a few millimetres from the colony's edge, and, there, each divides into several short bacilli—similar to those in the original colony; these cells grow and divide normally for a number of generations, forming a ring of heavy growth which surrounds (and is concentric with) the original colony. Later, another generation of swarm cells is formed at the outer edge of the ring and the cycle is repeated. In this way the entire

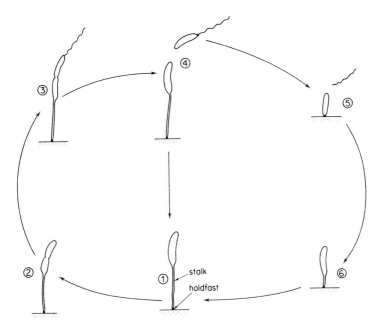

Fig. 4.1 The life-cycle of *Caulobacter*. 1. A mature, stalked cell attached to a surface by an adhesive 'holdfast'. 2. The stalked cell begins to divide. 3. A flagellum develops at the free end of the daughter cell. 4. Asymmetric binary fission is complete, and the motile daughter cell (*swarm cell*) swims away. The stalked cell can continue to grow and produce more swarm cells. 5. The swarm cell loses its flagellum and becomes attached by its holdfast. 6. A stalk develops, and the daughter cell matures into a new stalked mother cell.

surface of the medium becomes covered by concentric rings of growth. This phenomenon is called *swarming*.

Swarming is not essential for *Proteus*, and it doesn't happen on all types of medium. It occurs typically on rather moist surfaces; under such conditions, swarm cells help the organism to spread to new sources of nutrients.

4.3 RESTING CELLS

In some bacteria differentiation can result in the formation of a resting cell—either a *spore* or a *cyst*. Resting cells may function as disseminative units (helping to spread the organism) and/or as dormant cells which are capable of surviving in a hostile environment. Under suitable conditions a spore or cyst *germinates* to form a new vegetative cell.

4.3.1 Endospores

Endospores have been studied more thoroughly than any other type of bacterial spore: they are formed by species of *Bacillus*, *Clostridium*, *Coxiella*, *Desulfotomaculum*, *Thermoactinomyces* and a few other genera. An endospore is formed *within* a cell as a response to starvation—specifically, a shortage of carbon, nitrogen and/or phosphorus. It exists in a state of dormancy: few, if any, of the chemical reactions in a vegetative (growing) cell take place in the mature endospore. Not only can dormancy persist for long periods (estimated to be up to 1000 years for *Bacillus*), but the endospore is highly resistant to many hostile factors: extremes of temperature and pH, desiccation, radiation, various chemical agents and physical damage; in fact, irreversible inactivation of endospores can be guaranteed only by the harsh treatment of a *sterilization* process (section 15.1).

The formation of an endospore is shown diagrammatically in Fig. 4.2, and its structure is shown in Plate 3. At molecular level, the 'trigger' for endospore formation is widely believed to be a shortage of guanine nucleotides (or derivatives) (Figs 7.1–7.3) within the cell. The heat-resistance of the endospore is believed to be due to its low level of water; calcium dipicolinate (in the core) may act as a secondary stabilizer.

Under suitable conditions, an endospore *germinates*, i.e. it becomes metabolically active. The endospores of some species need to be 'activated' before they can germinate; activation may consist e.g. of sublethal heating or exposure to certain chemicals. Germination is promoted by chemicals called *germinants*; according to species, germinants may include e.g. L-alanine, some purine nucleosides, various ions or certain sugars. Germination may be initiated by the binding of a germinant to an inner membrane receptor. The transition from a germinated endospore to a vegetative cell is called *outgrowth*.

The genetic control of endospore formation and germination is considered briefly in section 7.8.2.3.

Note. 'Endospore' is often abbreviated to 'spore'. However, the endospore should not be confused with other types of bacterial spore (see section 4.3.2).

4.3.2 Other bacterial spores

In many of the hypha-forming actinomycetes, *exospores* are produced by septation and fragmentation of the hyphae (Fig. 4.3). These spores lack specialized structures (such as cortex and spore coat) but they do show some resistance e.g. to dry heat, desiccation and certain chemicals. The spores of *Streptomyces* are metabolically less active than the vegetative hyphae, though they are not completely dormant.

In the actinomycetes *Actinoplanes* and *Pilimelia*, motile (flagellated) *zoospores*

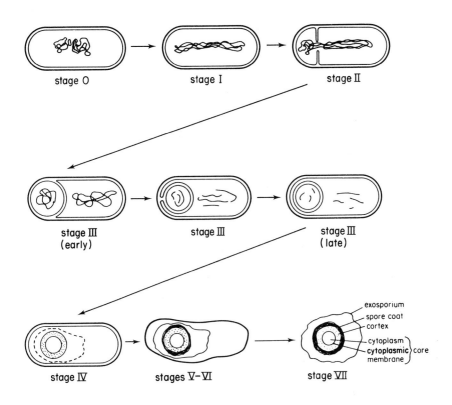

stage 0   stage I   stage II

stage III   stage III   stage III
(early)   (late)

exosporium
spore coat
cortex
cytoplasm
cytoplasmic } core
membrane

stage IV   stages V–VI   stage VII

Fig. 4.2 Endospore formation. Stage 0. A vegetative cell containing two chromosomes is about to sporulate. Stage I. An *axial filament* composed of the two chromosomes develops. Stage II. The cytoplasmic membrane grows inwards to form a septum which divides the protoplast into two unequal parts. Stage III. The septum is complete; the smaller of the two protoplasts will become the endospore and is called the *forespore* or *prespore*. The cytoplasmic membrane of the larger protoplast invaginates to engulf the forespore. When completely engulfed, the forespore is bounded by two membranes. Stage IV. A layer of modified peptidoglycan is laid down between the two membranes of the forespore to form a rigid layer called the *cortex*. A loose protein envelope called the *exosporium* may begin to develop at about this time. Stages V–VI. A multilayered protein *spore coat* is deposited outside the outmost membrane (stage V), and the spore matures (stage VI) to develop its characteristic resistance to heat and its bright, refractile appearance under the light microscope; during this time, calcium dipicolinate accumulates in the 'core'—i.e. the spore's protoplast (bounded by the inner membrane). Stage VII. The completed spore is released by disintegration of the mother cell. (An exosporium is not present on all endospores.)

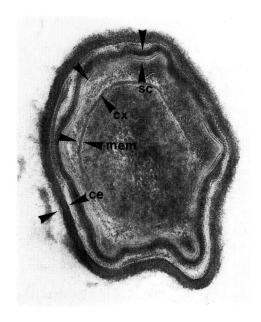

Plate 3. Cross-section of an endospore of *Bacillus subtilis* within the mother cell (approx. 1 µm across). The 'core' (protoplast) of the endospore is bounded by the membrane (mem). Between the membrane and the multilayered spore coat (sc) is the cortex (cx). Surrounding the endospore is the cell envelope (ce) of the mother cell.

Courtesy of Dr John Coote, University of Glasgow.

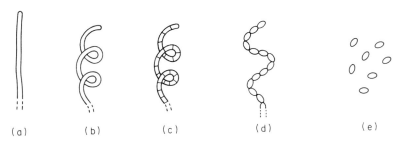

(a)   (b)   (c)   (d)   (e)

Fig. 4.3 Exospore formation in a species of *Streptomyces* (an actinomycete). (a) The tip of a vegetative aerial hypha. (b) The tip of the hypha becomes coiled. (c) Septa develop along the length of the coiled hypha. (d) The walls of the developing spores thicken, and each spore becomes rounded. (e) The spores are released.

are formed inside a closed sac called a *sporangium*; sporangia develop from the vegetative hyphae.

4.3.3 Bacterial cysts

Most studies on bacterial cysts have been carried out on the soil bacterium *Azotobacter vinelandii*. The desiccation-resistant cysts of this organism are dormant, and they can survive in dry soil for years. Encystment may be triggered by changes in the levels of carbon and nitrogen in the environment; it involves loss of flagella and the development of a complex cyst wall which contains alginate, protein and lipid. Typically, PHB (section 2.2.4.1) accumulates in the cyst.

4.4 AKINETES, HETEROCYSTS AND HORMOGONIA

These structures are formed by various filamentous (trichome-forming) *cyanobacteria*—photosynthetic organisms which occur e.g. in soil, in natural waters and in symbiotic associations with certain eukaryotes.

4.4.1 Akinetes

Akinetes are differentiated cells produced by some species e.g. under starvation conditions; each has a thickened cell wall and a cytoplasm rich in storage compounds (such as glycogen). Akinetes are usually larger than vegetative cells, and they have a lowered rate of metabolism; they show some resistance to desiccation and cold, and may function as overwintering and/or disseminative units.

4.4.2 Heterocysts

Heterocysts are formed by some species when there is a shortage of usable nitrogen compounds. Under such conditions, some of the cells in a trichome undergo differentiation, each forming a *heterocyst*: a specialized compartment within which atmospheric (gaseous) nitrogen can be 'fixed' —i.e. converted to a usable nitrogen compound (section 10.3.2.1). The process of differentiation includes e.g. development of a thick envelope, re-arrangement of the thylakoids, cessation of (photosynthetic) oxygen evolution, and synthesis of *nitrogenase* (an enzyme used in nitrogen fixation). Communication between a heterocyst and an adjacent vegetative cell occurs via fine pores (*microplasmodesmata*) in their contiguous cytoplasmic membranes; during nitrogen fixation, fixed nitrogen is transferred to the vegetative cell which, in turn, transfers carbon and other materials to the heterocyst.

4.4.3 Hormogonia

A hormogonium is a short trichome, lacking both akinetes and heterocysts, formed e.g. from a vegetative trichome; the cells of a hormogonium may be smaller than those of the parent trichome. Typically, hormogonia exhibit gliding motility. In some species (e.g. *Nostoc muscorum*) only the hormogonia contain gas vacuoles—reinforcing the idea that these short trichomes have a primarily disseminative role.

5 Metabolism I: energy

'Metabolism' refers to the chemical reactions which occur in living cells: molecules are built up (*anabolism*), broken down (*catabolism*) or changed from one type to another, and various atoms are oxidized or reduced. Most of the reactions involve specific protein catalysts called *enzymes*.

A sequence of metabolic reactions, in which one substance is converted to another (or others), is called a *metabolic pathway*; in such a pathway the *substrate* (e.g. a nutrient) is converted, often via one or more *intermediates*, to so-called *end-product(s)*. Many of the metabolic pathways in bacteria do not occur in eukaryotes.

Metabolic reactions are commonly *endergonic*, i.e. they require energy; energy is also needed for locomotion (in motile species) and for the uptake of various nutrients. Most bacteria obtain energy by 'processing' chemicals from the environment; such bacteria are called *chemotrophs*. Other bacteria, which use the energy in sunlight, are called *phototrophs*. However, neither chemicals from the environment nor sunlight can be used *directly* to fuel a cell's energy-requiring processes; consequently, the cell must have ways of converting 'environmental' sources of energy into a usable form of energy. What *is* a 'usable form of energy'? Given certain chemicals, or sunlight, cells can make specific 'high-energy' compounds with which they can satisfy their energy demands; these compounds include adenosine 5'-triphosphate (ATP), phosphoenolpyruvate (PEP), acetyl phosphate and acetyl-CoA. Such compounds have been called 'energy currency molecules' because the cell can spend them (rather than 'environmental' energy) just as we spend coins and banknotes instead of gold bars. Some 'currency' molecules are shown in Fig. 5.1.

ATP (Fig. 5.1a) yields usable energy when its terminal phosphate bond is broken; accordingly, as molecules of ATP are used up (supplying the cell's energy needs) molecules of adenosine 5'-diphosphate (ADP) are formed. Hence, environmental energy must be harnessed in such a way that ATP can be re-synthesized by the phosphorylation of ADP.

Another type of energy currency molecule—nicotinamide adenine dinucleotide (NAD)—carries energy in the form of 'reducing power': it accepts and yields energy by being (respectively) reduced and oxidized (Fig. 5.1b). Other carriers of reducing power include NAD phosphate (NADP) and flavin adenine dinucleotide (FAD).

Environmental energy can also be converted to an electrochemical form of energy; this consists of a gradient of ions (usually protons: H^+) between

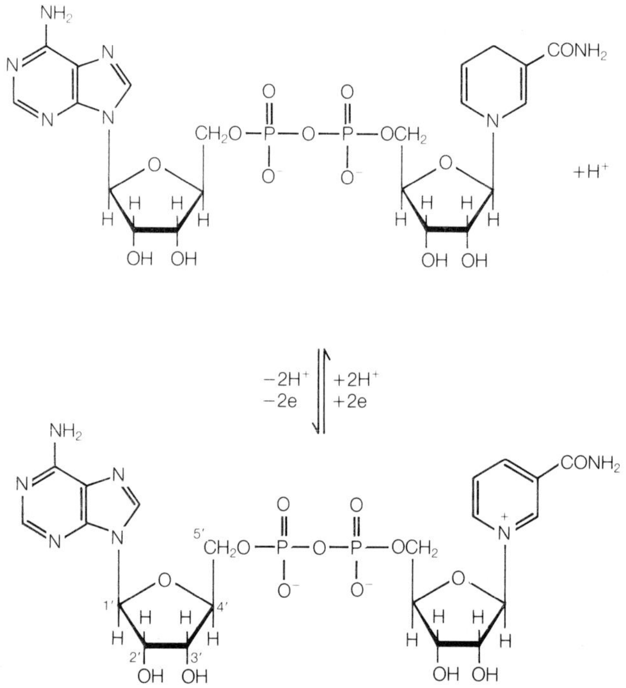

Fig. 5.1 Some energy currency molecules. (a) Adenosine 5'-triphosphate (ATP). When the γ-bond is broken the terminal phosphate group is lost, and the resulting molecule is adenosine 5'-diphosphate (ADP).

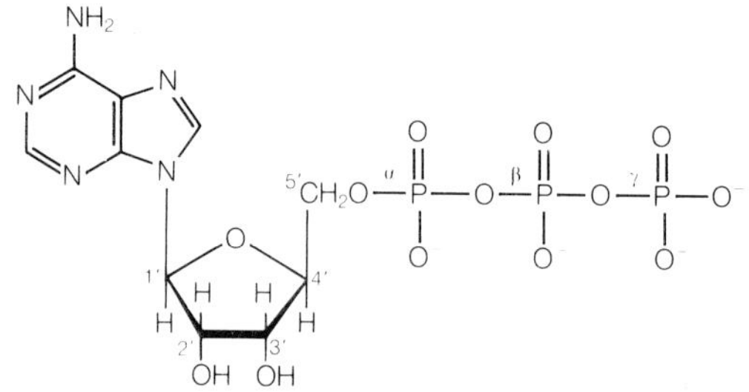

(b) Nicotinamide adenine dinucleotide (NAD) showing the reduced (upper) and oxidized (lower) forms; e = electron. Strictly, the oxidized form should be written NAD$^+$ but is often written NAD for convenience; similarly, the reduced form should be written NADH + H$^+$ but is often written NADH. NAD phosphate (NADP) is NAD 2'-phosphate—the phosphate group being at the 2'-position of the (left-hand) sugar (D-ribose) molecule.

the two surfaces of the cytoplasmic membrane. The energy in this ion gradient can be used for transport (see later), for driving flagellar rotation (section 2.2.14.1)—and for the synthesis of high-energy compounds!

How does a bacterium form high-energy compounds, or an ion gradient, from 'environmental' energy? Different strategies are used by chemotrophs and phototrophs, as discussed below.

5.1 ENERGY METABOLISM IN CHEMOTROPHS

The chemotrophs 'process' chemicals for energy; those which use organic compounds are called *chemoorganotrophs*, and those which use inorganic compounds, or elements, are called *chemolithotrophs*. (Mechanisms used for the *uptake* of chemicals are discussed in section 5.4.)

5.1.1 Energy metabolism in chemoorganotrophs

Chemoorganotrophs use their organic substrates in one of two main types of energy-converting metabolism: *fermentation* and *respiration*. Some chemoorganotrophs can carry out only one of these processes; others can carry out either—depending on conditions.

5.1.1.1 Fermentation

Fermentation is a type of energy-converting metabolism in which the substrate is metabolized *without the involvement of an exogenous (i.e. external) oxidizing agent*. (Note. Fermentation typically—but not necessarily—occurs anaerobically, i.e. in the absence of oxygen, but this is *not* the distinguishing feature of fermentation: as we shall see later, respiration also can occur anaerobically.) Because no external oxidizing agent is used, the products of fermentation—collectively—are neither more nor less oxidized than the substrate; that is, the oxidation of any intermediate in a fermentation pathway is balanced by the reduction of other intermediate(s) in that pathway. This is illustrated diagrammatically in Fig. 5.2.

These ideas can be understood more easily by looking at some real metabolic pathways. In many bacteria the fermentation of glucose begins with a pathway known as *glycolysis* or as the *Embden-Meyerhof-Parnas pathway* (EMP pathway) (Fig. 5.3). In this pathway, 1 molecule of glucose yields (via a number of intermediates) 2 molecules of the end-product, pyruvic acid. At two places in this pathway (Fig. 5.3) energy from *exergonic* (i.e. energy-yielding) reactions is used to phosphorylate ADP—that is, to synthesize the 'energy currency molecule' ATP from ADP. When energy from a chemical

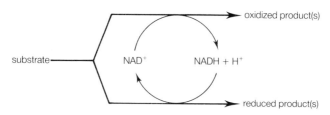

Fig. 5.2 A diagrammatic view of a fermentation pathway. Here, the (organic) substrate gives rise to intermediates which undergo *mutual* oxidation and reduction; taken together, the products have the same oxidation state as the original substrate. (Compare with Fig. 5.4.)

reaction is used, directly, for the synthesis of ATP from ADP the process is known as *substrate-level phosphorylation*.

In terms of energy currency molecules, the EMP pathway can be summarized as follows:

$$2ATP + 4ADP + 2NAD \rightarrow 2ADP + 4ATP + 2NADH$$

Thus, each molecule of glucose metabolized gives 2NADH and a *net* yield of 2ATP. For the metabolism of further molecules of glucose there is clearly a need for fresh supplies of both ADP and NAD. ADP is regenerated from ATP when the latter is used to supply energy. However, as no external oxidizing agent is used in fermentation, how is NAD regenerated from NADH? Earlier it was said that the fermentation of glucose may *begin* with the EMP pathway; in fact, the EMP pathway is only the 'front end' of a number of different pathways: what happens to the NADH (and pyruvic acid) depends on subsequent reactions. In the simplest case, NADH donates its reducing power to the pyruvic acid, i.e. the oxidation of NADH to NAD is coupled with the reduction of pyruvic acid to lactic acid:

$$\text{pyruvic acid} + NADH \rightarrow \text{lactic acid} + NAD$$

This reaction completes one possible fermentation pathway: a so-called *lactic acid fermentation*; a pathway in which lactic acid is the only (or predominant) product is called a *homolactic fermentation*. Lactic acid—a waste product—is released by the cells.

Notice that lactic acid ($C_3H_6O_3$) has the same oxidation state as glucose ($C_6H_{12}O_6$)—i.e. no net oxidation or reduction has occurred; although glyceraldehyde 3-phosphate undergoes oxidation, this is balanced by the reduction of pyruvic acid to lactic acid (Fig. 5.4).

Other fermentation pathways which have the EMP pathway as their 'front end' include the mixed acid fermentation and the butanediol fermentation. In these pathways the pyruvic acid is metabolized to several end-products, the relative proportions of which can vary according to conditions of growth.

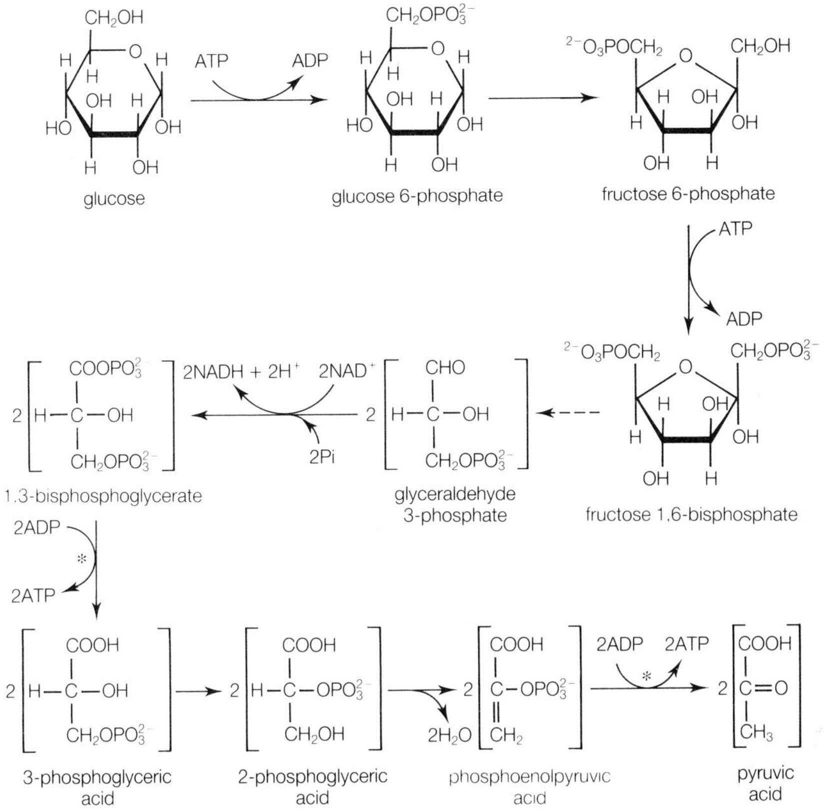

Fig. 5.3 The Embden–Meyerhof–Parnas pathway. The broken arrow from fructose 1,6-bisphosphate indicates a simplified section of the pathway. Each of the two substrate-level phosphorylations is marked with an asterisk; in each case, phosphate is transferred to ADP from an energy-rich organic phosphate. Although two reactions in the pathway actually require ATP, there is nevertheless a *net* gain of 2ATP for each molecule of glucose metabolized. (*Note*. Pyruvic acid and phosphoenolpyruvic acid are often referred to as 'pyruvate' and 'phosphoenol-pyruvate', respectively.)

The *mixed acid fermentation* (Fig. 5.5) occurs e.g. in *Escherichia coli* and in species of *Proteus* and *Salmonella*. Under acidic conditions, *E. coli*, and other species which have the appropriate enzyme system, can split the formic acid into carbon dioxide and hydrogen; thus, these organisms carry out an *aerogenic* (gas-producing) fermentation. Organisms which lack the enzyme system carry out the fermentation *anaerogenically* (i.e. without forming gas); they include species of *Shigella*.

The *butanediol fermentation* (Fig. 5.6) occurs e.g. in species of *Enterobacter*, *Erwinia*, *Klebsiella* and *Serratia*. The amount of acid formed in this

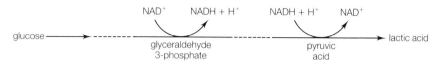

Fig. 5.4 Homolactic fermentation: a diagrammatic view of the fermentation of glucose to lactic acid via the Embden–Meyerhof–Parnas pathway. Note that the oxidation is balanced by an equivalent reduction. The homolactic fermentation is carried out by some of the so-called 'lactic acid bacteria' which are used in the food industry.

fermentation is generally much less than that formed in the mixed acid fermentation. Some strains form small amounts of diacetyl $(CH_3.CO.CO.CH_3)$ from the acetolactic acid under certain conditions.

Although the mixed acid and butanediol fermentations are more complicated than the homolactic fermentation, they nevertheless conform to the same basic principles; while the relative proportions of end-products may vary, the formation of products more oxidized than glucose (e.g. formic acid) is always balanced by the formation of products more reduced than glucose (e.g. ethanol).

NADH and other compounds formed during fermentation can be used in various ways. Some of the NADH will be oxidized in biosynthetic reactions (rather than in the formation of waste products such as lactic acid); this also allows compounds such as pyruvate to be used as precursor molecules for biosynthesis (section 6.3.1 and Fig. 6.3). Both in respiration (section 5.1.1.2) and fermentation, there is (in chemoorganotrophs) a close link between energy metabolism and carbon metabolism (section 5.3.1).

Note. Before leaving the topic of fermentation it is worth mentioning that, in industry, the term 'fermentation' is sometimes used for *any* chemical process mediated by microorganisms—even for those processes in which fermentation is not involved.

5.1.1.2   *Respiration*

Respiration is a type of energy-converting metabolism in which the substrate is metabolized *with the involvement of an exogenous (i.e. external) oxidizing agent* (compare fermentation: section 5.1.1.1). Respiration can occur in the presence of oxygen (oxygen itself being the external oxidizing agent)—but it can also occur anaerobically (when other inorganic or organic oxidizing agents are used in place of oxygen). (As we are still talking about chemoorganotrophs, the *substrate* is always an organic compound, even though the oxidizing agent may be inorganic or organic.)

Because an external oxidizing agent is used, the substrate undergoes a *net*

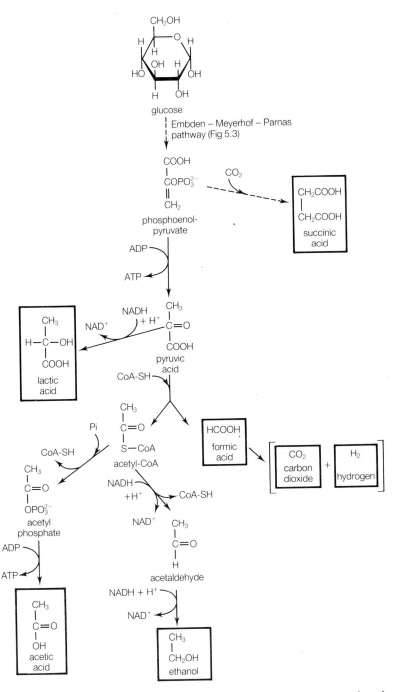

Fig. 5.5 The mixed acid fermentation. CoA = coenzyme A; Pi = inorganic phosphate.

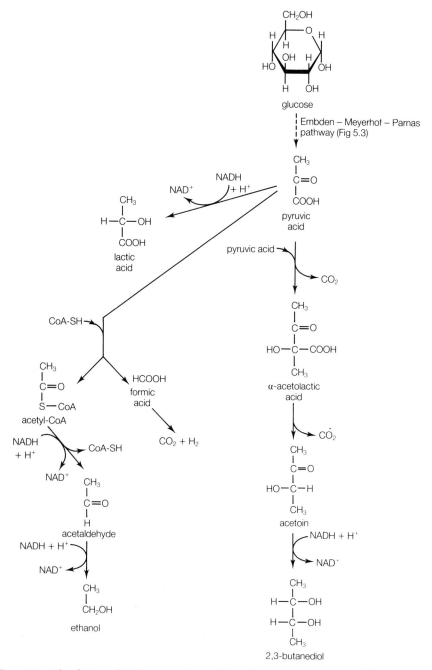

Fig. 5.6 The butanediol fermentation. (See also section 16.1.2.5—the Voges–Proskauer test.)

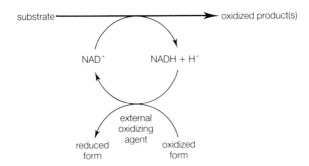

Fig. 5.7 Respiration (diagrammatic): the relationship between substrate, end-product(s) and external oxidizing agent in the respiration of a typical *organic* substrate (compare Fig. 5.2). In this example, NADH—formed during metabolism of the substrate—is oxidized by the external oxidizing agent; NADH and the external oxidizing agent are generally coupled indirectly via an electron transport chain (see text).

oxidation (Fig. 5.7); glucose, for example, can be oxidized to carbon dioxide and water. The oxidation of a substrate provides more energy than that obtainable—from the same substrate—by fermentation.

How is the substrate oxidized, and how is usable energy obtained? The usual mode of oxidation of a typical *organic* substrate is shown in Fig. 5.7: oxidation is coupled with the reduction of NAD, the resulting NADH being oxidized by an external oxidizing agent. NADH and the external oxidizing agent usually interact indirectly via an *electron transport chain* (ETC) located in the cytoplasmic membrane. An ETC is a chain of specialized molecules (redox agents) which form a conducting path for electrons; the sequence of the (different) redox agents is such that electrons can flow down a redox gradient (towards the more positive end) in a series of oxidation/reduction reactions. Electrons from NADH flow down the gradient to an external oxidizing agent; when the latter is oxygen, the situation can be summarized as in Fig. 5.8. The final recipient of the electrons (in this case oxygen) is called the *terminal electron acceptor*.

Electron flow of this kind necessarily yields energy because the electrons are moving from high-energy to lower-energy locations. Typically, this liberated energy causes protons (hydrogen ions: H^+) to be pumped across the cytoplasmic membrane—from the inner to the outer surface; this creates an imbalance of electrical charge (and pH) between the two surfaces of the membrane. The tendency of protons to move back across the membrane (and thus abolish the imbalance) constitutes a form of energy known as *proton motive force*. Proton motive force (pmf) is one of the cell's most important and versatile forms of energy. It can be used—directly—to satisfy several types of energy demand. Thus, it can drive flagellar rotation (section 2.2.14.1). It can provide energy for the transport (uptake) of various ions;

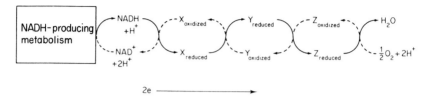

Fig. 5.8   Respiration (diagrammatic): an electron transport chain (ETC) consisting of three components (X, Y and Z), with oxygen as the terminal electron acceptor, in the respiration of a typical *organic* substrate. The cell's ongoing metabolism produces NADH, from which NAD must be regenerated by oxidation; some NADH is oxidized to NAD in biosynthetic reactions, and some is oxidized via the ETC. In the ETC, NADH is oxidized by transferring electrons to (and thus reducing) the oxidized form of X. The reduced form of X is oxidized by transferring electrons to the oxidized form of Y—and so on. The final step is the reduction of oxygen to water. The solid curved lines indicate the path of electron flow. Oxidation of NADH (enzyme: NADH dehydrogenase) and reduction of oxygen (at a cytochrome oxidase) both appear to occur at the *inner* (i.e. cytoplasmic) face of the cytoplasmic membrane.

ion uptake is an energy-requiring process because the cytoplasmic membrane is ordinarily impermeable to ions (section 2.2.8). It can provide energy for the transport of certain substrates across the cytoplasmic membrane—e.g. lactose uptake in *E. coli*. Pmf can also provide energy for the phosphorylation of ADP to ATP at enzyme complexes (*ATPases*) located in the cytoplasmic membrane. In respiration, when pmf is used as a source of energy for the synthesis of ATP from ADP the process is called *oxidative phosphorylation* (compare 'substrate-level phosphorylation': section 5.1.1.1). Interestingly, membrane ATPases can also catalyse the hydrolysis of ATP to ADP, the liberated energy being used to pump protons across the membrane—i.e. to augment pmf; thus, the energy in ATP and pmf is interconvertible! These ideas are summarized in Fig. 5.9.

In a bacterium, the electron transport chain involved in respiration (the *respiratory chain*) occurs in the cytoplasmic membrane; its components vary from species to species, and variations occur even in a given species growing under different conditions. Components found in respiratory chains include: (i) *cytochromes*: iron-containing proteins which receive and transfer electrons by the alternate reduction and oxidation of the iron atom; (ii) *iron–sulphur proteins* such as the *ferredoxins*; and (iii) *quinones*: aromatic compounds which can undergo reversible reduction. In the respiration of a typical organic substrate (Fig. 5.8), the oxidation of NADH and the reduction of the terminal electron acceptor (oxygen in Fig. 5.8) both appear to occur at the *inner* (i.e. cytoplasmic) face of the cytoplasmic membrane.

In Fig. 5.8 the source of NADH is given simply as 'NADH-producing metabolism'. We can now look at some actual pathways used in bacterial respiration. In many bacteria, the respiratory metabolism of glucose starts with the Embden–Meyerhof–Parnas pathway (EMP pathway, Fig. 5.3).

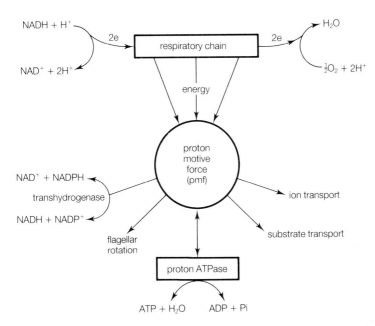

Fig. 5.9 Some of the roles of proton motive force. 'Respiratory chain' is the term used for an electron transport chain involved in respiratory metabolism (i.e. respiration). 'Proton ATPase' is an enzyme system (in the cytoplasmic membrane) which catalyses the pmf-dependent phosphorylation of ADP to ATP as well as the hydrolysis of ATP to ADP and inorganic phosphate (Pi); pmf is *used* for the synthesis of ATP but is *augmented* by the (energy-yielding) hydrolysis of ATP. Pmf also controls the (reversible) reduction of NADP by NADH; NADPH is used e.g. for some of the cell's biosynthetic reactions.

Beyond pyruvic acid, however, the pathways of fermentation and respiration are completely different. In respiration, pyruvic acid is often converted to acetyl-CoA and fed into a cyclical pathway known as the *tricarboxylic acid cycle* (TCA cycle, Fig. 5.10)—also known as the *Krebs cycle* or the *citric acid cycle*.

Figure 5.10 shows that, in the TCA cycle, acetyl-CoA and oxaloacetic acid (OAA) combine to form citric acid; in the subsequent reactions, the original molecule of pyruvic acid is, in effect, oxidized to carbon dioxide. For each molecule of pyruvic acid oxidized, 4 molecules of NAD(P) and 1 of FAD are reduced, 1 molecule of ATP is synthesized, and 1 molecule of OAA is regenerated. NADH and $FADH_2$ can be oxidized via a respiratory chain, the resulting pmf being used e.g. for the synthesis of ATP at a membrane ATPase. In terms of energy yield, it should now be clear that respiration is much more efficient than fermentation: in respiration, the oxidation of NADH can lead, via pmf, to the synthesis of ATP, whereas in fermentation (where there is no external oxidizing agent) the cell has to get rid of NADH by synthesizing waste products such as lactic acid.

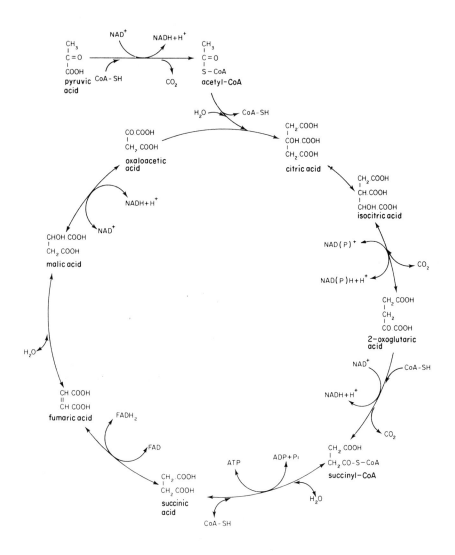

Fig. 5.10 The tricarboxylic acid cycle (TCA cycle): a common pathway in respiratory metabolism (section 5.1.1.2). The reaction isocitric acid → 2-oxoglutaric acid is catalysed by the enzyme isocitrate dehydrogenase which, in most bacteria, is specific for NADP rather than NAD; NADPH is used e.g. in biosynthetic reactions. FAD (flavin adenine dinucleotide)—like NAD—carries energy in the form of 'reducing power'; $FADH_2$ can be oxidized via a respiratory chain, oxidation yielding pmf. In many bacteria the step from succinyl-CoA to succinic acid involves the synthesis of guanosine 5'-triphosphate (GTP) rather than ATP; GTP is used as an energy currency molecule e.g. in the binding of the aminoacyl-tRNA to the 'A' site in protein synthesis (Fig. 7.9).

Anaerobic respiration. In principle, anaerobic respiration is the same as aerobic respiration: both use an external oxidizing agent. Under anaerobic conditions, however, oxidizing agents such as nitrate, sulphate and fumarate are used in place of oxygen; pmf can be generated by an anaerobic respiratory chain.

In *nitrate respiration* nitrate is used as the terminal electron acceptor, the nitrate being reduced to nitrite, nitrous oxide, nitrogen or ammonia—depending on species. When the nitrate is reduced mainly to nitrogen and/or nitrous oxide (i.e. gases) the process is called *denitrification*; this process is important agriculturally: it can lower soil fertility by causing a loss of nitrogen (section 10.3.2.2). Bacteria capable of denitrification include *Bacillus licheniformis, Paracoccus denitrificans* and *Pseudomonas stutzeri*; some of the so-called 'denitrifying bacteria' can be useful for eliminating nitrate from waste water.

The *sulphate-* and *sulphur-reducing bacteria* use sulphate and sulphur, respectively, as terminal electron acceptors; during anaerobic respiration, sulphate or sulphur is reduced to sulphide (Fig. 10.3). This type of anaerobic respiration is carried out e.g. by species of *Desulfococcus* and *Desulfovibrio* (sulphate reducers) and *Desulfuromonas* (sulphur reducer); these organisms typically occur in anaerobic mud and soil, and they are responsible for much of the hydrogen sulphide found in organically polluted waters.

In *fumarate respiration* the terminal electron acceptor is exogenous fumarate—which is reduced to succinate. Fumarate respiration is carried out by a range of bacteria, including *E. coli*, under appropriate conditions.

5.1.2 Energy metabolism in chemolithotrophs

Chemolithotrophs use *inorganic* substrates for energy metabolism—e.g. sulphide, elemental sulphur, ammonia, hydrogen, ferrous ions etc. Metabolism usually involves aerobic or anaerobic respiration: electrons from the substrate are transferred to the external oxidizing agent, pmf being generated (see also section 5.3.4); NAD is not involved (compare respiration in chemoorganotrophs). Obligate or facultative chemolithotrophs include (for example) the thiobacilli, the nitrifying bacteria and certain methanogens.

Species of *Thiobacillus* occur e.g. in soil and marine muds. Typically, they oxidize substrates such as sulphide and sulphur aerobically, though *T. denitrificans* can metabolize these substrates anaerobically by nitrate respiration. *T. ferrooxidans*, a strict aerobe, can oxidize ferrous ions as well as sulphur substrates. Thiobacilli are important in the sulphur cycle (section 10.3.3).

Nitrifying bacteria are obligately aerobic respiratory organisms which live in soil and aquatic environments. They obtain energy by *nitrification*: the oxidation of ammonia to nitrite (e.g. species of *Nitrosococcus* and *Nitrosomonas*) and nitrite to nitrate (e.g. species of *Nitrobacter* and *Nitrococcus*). These bacteria are important in the nitrogen cycle (section 10.3.2).

Methanogens are obligately anaerobic archaebacteria which live e.g. in river muds and in the rumen in cows and other ruminants. Many methanogens (e.g. species of *Methanobacterium, Methanobrevibacter*) can obtain energy by a complex series of reactions in which both hydrogen and carbon dioxide are metabolized and methane is produced as a waste product. (Some species form methane from e.g. acetate or methanol.)

5.1.2.1   Inorganic fermentation

Until quite recently it was thought that no *inorganic* substrate could be fermented. Then, in 1987, Bak and Cypionka [Nature (1987) *326* 891–892] described a type of energy metabolism in which the substrate (sulphite *or* thiosulphate) underwent 'disproportionation' to yield sulphide and sulphate. This 'chemolithotrophic fermentation' occurs e.g. in *Desulfovibrio sulfodismutans.*

5.2   ENERGY METABOLISM IN PHOTOTROPHS

Phototrophs obtain energy from sunlight—in most cases by *photosynthesis.*

5.2.1   Photosynthesis

In photosynthesis, the energy in light is absorbed by specialized pigments and is used to form energy currency molecules and/or pmf; in all cases, photosynthesis occurs in membranes containing *chlorophylls*, accessory pigments and electron transport chain(s). Chlorophylls are green, magnesium-containing pigments. Cyanobacteria contain chlorophyll *a* (which also occurs in algae and higher plants); the other photosynthetic bacteria contain one or more *bacteriochlorophylls*—pigments which are similar to the chlorophylls. The so-called *light reaction* of photosynthesis refers to all the (photochemical) events involved in the conversion of light energy to pmf or chemical energy; the *dark reaction* (= light-independent reaction) refers to the cell's use of its photosynthetically-derived energy for the synthesis of carbon compounds.

Photosynthetic bacteria can be divided into two main categories: (i) those which carry out photosynthesis aerobically (and which produce oxygen as a by-product), and (ii) those which carry out photosynthesis anaerobically (and which do not produce oxygen).

5.2.1.1   Oxygenic (oxygen-producing) photosynthesis in bacteria

This process—which closely resembles photosynthesis in green plants and algae—occurs in the cyanobacteria. Because the cyanobacteria carry out eukaryotic-type photosynthesis they were, for many years, regarded not as bacteria but as 'blue–green algae'; today, the bacterial nature of these organisms is not in doubt.

In almost all cyanobacteria photosynthesis occurs in the thylakoid membranes (section 2.2.7). (In strains of *Gloeobacter*, which lack thylakoids, the photosynthetic components appear to occur in the cytoplasmic membrane.) Chlorophylls occur in so-called *reaction centres* to which light is channelled by specialized protein–pigment *light-harvesting complexes*. When the chlorophyll receives light energy it ejects highly energized electrons; these electrons can flow down an electron transport chain and provide energy for (i) pmf generation and/or (ii) the direct reduction of NADP. (The cell uses NADPH e.g. for various biosynthetic reactions.)

Oxygenic photosynthesis is generally represented by the *Z scheme* (Fig. 5.11). Energized electrons ejected from photosystem II (PSII) flow down an electron transport chain to photosystem I (PSI), and this electron flow generates pmf across the thylakoid membrane; when such photosynthetically-derived pmf is used for the synthesis of ATP (at a membrane-bound ATPase) the process is called *photophosphorylation.* Electrons ejected from PSI have sufficient energy to reduce NADP to NADPH. The flow of electrons (from left to right in Fig. 5.11) requires an input of electrons to PSII. This is achieved by the oxidation of water—oxygen being liberated as a by-product.

Some cyanobacteria (e.g. *Oscillatoria limnetica*) can—as an alternative—carry out photosynthesis anaerobically, using sulphide instead of water as an electron donor; the sulphide is oxidized to elemental sulphur. Some can even grow as chemoorganotrophs.

5.2.1.2   Anoxygenic photosynthesis

Anoxygenic photosynthesis (in which oxygen is not produced) is carried out anaerobically by bacteria of the order Rhodospirillales. In the so-called 'purple' photosynthetic bacteria (suborder Rhodospirillineae) all the photosynthetic components occur in intracellular membranes which are continuous with the cytoplasmic membrane. In the 'green' photosynthetic bacteria (suborder Chlorobiineae) components of the light-harvesting complexes occur in *chlorosomes* (section 2.2.7) while the reaction centres occur in the cytoplasmic membrane.

In the 'purple' bacteria, electrons ejected from a reaction centre follow a *cyclic* path—via an electron transport chain—back to the reaction centre; the pmf which is generated can be used e.g. for the synthesis of ATP (i.e. photophosphorylation).

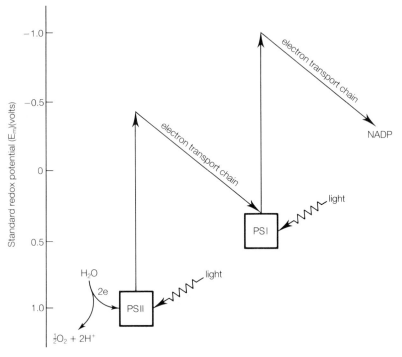

Fig. 5.11   Oxygenic photosynthesis: simplified Z scheme (scale approximate). Light energy causes the reaction centres (PSI and PSII) to eject energized electrons; the energy levels and destinations of these electrons are shown by the arrows. Electrons ejected from PSII are replaced by the oxidation of water, oxygen being liberated. The flow of electrons from PSII to PSI generates pmf.

In 'green' bacteria the electron flow generates pmf (for e.g. ATP synthesis) and it can also be used for the direct reduction of NAD to NADH. Such *non-cyclic* electron flow needs a supply of electrons from an exogenous donor; electron donors used by the 'green' photosynthetic bacteria include e.g. sulphide and thiosulphate—but never water, because in these organisms photosynthesis is always anoxygenic.

Some bacteria of the Rhodospirillales (including many 'purple' bacteria) can live as chemoorganotrophs under aerobic or microaerobic conditions.

5.2.1.3   Electron donors in photosynthesis

Inorganic electron donors in photosynthesis include e.g. water (used only by cyanobacteria), sulphide, sulphur and hydrogen; organic donors include e.g. formic acid and methanol. Phototrophs which use inorganic electron donors are called *photolithotrophs*; those which use organic donors are called *photoorganotrophs*.

5.2.2 The purple membrane

Certain strains of *Halobacterium salinarium* (an extreme halophile—section 3.1.7) can use the energy in sunlight even though they cannot carry out photosynthesis. In these strains, a differentiated, purple-pigmented region of the cytoplasmic membrane—the 'purple membrane'—develops under anaerobic or microaerobic conditions in the light; the main purple pigment is called *bacteriorhodopsin*. When a molecule of bacteriorhodopsin absorbs light energy it passes rapidly through a cycle of states (photointermediates), all within a few thousandths of a second; during this time protons are pumped outwards across the membrane, i.e. the 'photocycling' in bacteriorhodopsin generates pmf. Unlike photosynthesis, pmf generation appears not to involve electron transport.

5.3 OTHER ASPECTS OF ENERGY METABOLISM

5.3.1 The yield of ATP

In a living cell, no pathway or process occurs in isolation but rather forms part of a complex whole. Because of this, the maximum theoretical yield of ATP from the respiration of (say) one molecule of glucose may not be achieved owing to interactions between the respiratory pathway and other pathways in the cell. For example, not all the energy liberated (as pmf) by electron transport may be used for oxidative phosphorylation: some may be needed e.g. for ion transport or flagellar motility. Again, some of the NADH generated during respiration may be used for biosynthesis; this NADH is clearly not available for oxidation (via the respiratory chain) so that it cannot contribute to pmf and, hence, ATP synthesis. Finally, intermediates in energy metabolism may be drawn off to supply the cell with 'building blocks' for biosynthesis; thus, e.g. pyruvic acid is used in the synthesis of the amino acids alanine, valine and leucine. The withdrawal of intermediates necessarily sacrifices the energy which would otherwise have been obtained by their metabolism.

5.3.2 Reverse electron transport

Bacteria need 'reducing power'—e.g. NADH and/or NADPH—for biosynthesis. For fermentative bacteria this is never a problem, while the cyanobacteria and 'green' photosynthetic bacteria obtain these reduced molecules by direct reduction using non-cyclic electron flow. However, the 'purple' photosynthetic bacteria cannot do this: electrons ejected from their reaction centres do not have enough energy to reduce NAD. Instead, these organisms use *reverse electron transport*; in this process, the energy in pmf is used to drive electrons 'uphill' to a membrane-bound enzyme, NAD dehydrogenase, where NAD is reduced to NADH. This requires an input of electrons from an external electron donor; the 'purple' bacteria typically use organic electron donors, but some can use inorganic donors such as sulphide. Reverse electron transport is also used by the nitrifying bacteria and other chemolithotrophs; note that chemolithotrophs (section 5.1.2) do not form NADH during metabolism of the energy substrate.

5.3.3   End-product efflux

Fermentative bacteria make a great deal of NADH; in fact, they have to get rid of some of it by synthesizing waste products such as lactic acid (section 5.1.1.1). Making a virtue of necessity, some bacteria (e.g. *Streptococcus cremoris*) gain energy by linking protons to their waste lactate; when lactic acid passes outwards across the cytoplasmic membrane its 'passenger' protons automatically augment pmf!

5.3.4   Extracytoplasmic oxidation

Complex energy substrates are generally transported across the cytoplasmic membrane before being metabolized. However, some simple energy substrates (such as H_2 and Fe^{2+}) appear to be oxidized at the *outer* face of the cytoplasmic membrane or in the periplasmic region; such *extracytoplasmic oxidation* of a substrate is characterized by: (i) release of protons from the substrate and/or water extracytoplasmically, (ii) a transmembrane flow of electrons (from the substrate) to the cytoplasmic (inner) face of the membrane, and (iii) interaction of these electrons with protons and a terminal electron acceptor (e.g. oxygen). The net result is pmf generation. For example, the strict aerobe *Thiobacillus ferrooxidans* gains energy from the oxidation of Fe^{2+} (to Fe^{3+}) at low pH; according to one scheme, Fe^{3+} (produced extracytoplasmically) reacts with water to yield ferric hydroxide and protons, while electrons (from Fe^{2+}) reduce a terminal electron acceptor (oxygen?) at the cytoplasmic face of the membrane.

Some bacteria (e.g. *Pseudomonas aeruginosa*) can generate pmf by the extracytoplasmic oxidation of glucose to gluconate; in these bacteria the enzyme glucose dehydrogenase—with its cofactor *pyrroloquinoline quinone* (PQQ)—occurs bound to the outer surface of the cytoplasmic membrane. The gluconate can be transported across the membrane and phosphorylated to 6-phosphogluconate, which can enter the Entner–Doudoroff pathway (Fig. 6.2).

E. coli (and many related bacteria) normally form a membrane-bound glucose dehydrogenase which lacks the necessary PQQ; such organisms can carry out extracytoplasmic oxidation only if they are provided with PQQ

[Bouvet, Lenormand & Grimont (1989) IJSB *39* 61–67]. However, a mutant strain of *E. coli* with a non-functional PTS system (section 5.4) can synthesize PQQ and carry out extracytoplasmic oxidation—suggesting that PQQ-encoding genes may be present but not expressed under normal conditions [Biville, Turlin & Gasser (1991) JGM *137* 1775–1782].

5.4   TRANSPORT SYSTEMS

The cytoplasmic membrane (section 2.2.8) is an efficient barrier which stops most molecules (and all ions) from passing freely into and out of the cytoplasm. (This, of course, is necessary to allow the cell to control its own internal environment.) However, during ongoing metabolism the cell must be able to take up various substrates (for energy and other purposes) and get rid of waste products; this is the job of the various transport systems. Each transport system is typically specific for one or a few substrates, though a given substrate may have more than one type of transport system, even in the same cell. Transport often requires energy—energy being supplied e.g. by pmf or by a 'high-energy phosphate'.

Some transport systems are complex and are not yet fully understood; a few are outlined below.

Pmf can be used, directly, for the transport of certain charged and uncharged solutes. In *E. coli*, for example, 'proton–lactose symport' allows the uptake of lactose at the expense of pmf, lactose and protons being *jointly* transported into the cytoplasm.

Energy for the transport of ions is sometimes provided by the hydrolysis of ATP at a membrane-bound ATPase. For example, in *E. coli*, K^+ can be transported by the so-called 'Kdp' system which involves a specialized K^+-ATPase (or 'potassium pump'); hydrolysis of ATP at the 'pump' allows uptake of K^+ against a concentration gradient.

In binding-protein-dependent transport, the (extracellular) substrate is bound by a substrate-specific 'binding protein' in the periplasmic region; the bound substrate is then transferred to a protein complex in the cytoplasmic membrane which brings about transmembrane translocation of the substrate.

The phosphoenolpyruvate-dependent phosphotransferase system (mercifully written 'PTS system') is used by some Gram-positive and Gram-negative bacteria for the uptake of various sugars and other substrates. In this process, phosphoenolpyruvate (PEP) commonly supplies phosphate and energy for the phosphorylation of certain protein(s) in the cytoplasm; the phosphate (and energy) is handed on to proteins in the cytoplasmic membrane and, finally, to the substrate. Phosphorylation of the substrate is, in itself, an integral part of the transport process. Although pmf does not supply energy to this process, the *level* of pmf influences PTS activity.

Transport systems are found not only in the cytoplasmic membrane but in other membranes—such as the outer membrane in Gram-negative bacteria (section 2.2.9.2)—through which essential solutes must pass.

6 Metabolism II: carbon

Why carbon? Simply because virtually all the compounds which comprise—
and which are formed by—living organisms are carbon compounds. In this
chapter we consider (i) the carbon requirements of bacteria, (ii) the ways in
which different carbon compounds are assimilated, and (iii) the synthesis,
interconversion and polymerization of carbon compounds. The metabolism
of nitrogen and sulphur is considered in Chapter 10.

What sort of carbon compounds do bacteria need for growth? Some
bacteria are able to use carbon dioxide for most or all of their carbon
requirements; such bacteria are called *autotrophs*, and the use of carbon
dioxide as the sole (or main) source of carbon is called *autotrophy*. In some
autotrophic bacteria autotrophy is optional; in others it is obligatory: *only*
carbon dioxide can be used as a source of carbon—even when glucose or
other substrates are freely available. The *obligate* autotrophs use simple
energy substrates (Chapter 5) as well as a simple carbon source: they are
either chemolithotrophs (i.e. *chemolithoautotrophs*) or photolithotrophs (i.e.
photolithoautotrophs).

Chemolithoautotrophic metabolism occurs in relatively few bacteria
(which include some archaebacteria and some eubacteria); this type of
metabolism is unique in the living world, being found only in these
specialized bacteria and in no other type of organism. The chemolithoautotrophs have important roles in the cycles of matter (Chapter 10).

Photolithoautotrophic metabolism is found e.g. in the cyanobacteria—
and, of course, in green plants and algae.

Most bacteria are *not* autotrophic: they cannot use carbon dioxide as a
major source of carbon, and their growth depends on a supply of complex
carbon compounds derived from other organisms; bacteria which need
complex carbon compounds are called *heterotrophs*. Chemoorganotrophic
heterotrophs are *chemoorganoheterotrophs*. Collectively, the heterotrophs can
use a vast range of carbon sources—including sugars, fatty acids, alcohols
and various other organic substances. Heterotrophic bacteria are widespread
in nature, and they include (for example) all those species which cause
disease in man, other animals and plants.

6.1 CARBON ASSIMILATION IN AUTOTROPHS

In an autotroph, carbon dioxide from the environment is used to form complex organic compounds; when carbon dioxide is incorporated into such compounds it is said to have been 'fixed'. Different autotrophs have different pathways for carbon dioxide fixation, but there are two very common pathways: the Calvin cycle and the reductive TCA cycle.

6.1.1   The Calvin cycle

This pathway (also called the *reductive pentose phosphate cycle*) is used by a wide range of autotrophs, including some anoxygenic photosynthetic bacteria and most or all cyanobacteria. Part of the Calvin cycle is shown in Fig. 6.1. Each turn of the Calvin cycle requires a considerable input of both ATP and reducing power, i.e. carbon dioxide fixation needs a lot of energy. The key enzymes in this pathway are ribulose 1,5-bisphosphate carboxylase-oxygenase (RuBisCO—see also section 2.2.6) and phosphoribulokinase; these enzymes are found only in the Calvin cycle.

6.1.2   The reductive TCA cycle

This pathway is used for carbon dioxide fixation e.g. by phototrophic eubacteria of the family Chlorobiaceae, and by the chemotrophic archaebacterium *Sulfolobus*. Essentially, the pathway resembles the TCA cycle (Fig. 5.10) operating in reverse, with its one-way reactions (such as oxaloacetic acid → citric acid) being modified by different enzymes/reaction sequences. As in the Calvin cycle, carbon dioxide fixation requires a great deal of energy.

6.1.3   Carboxydobacteria

Carboxydobacteria can use carbon *monoxide* as the sole source of carbon and energy, i.e. they do not conform to the strict definition of 'autotroph'. However, they oxidize carbon monoxide to carbon dioxide, aerobically, and they assimilate carbon dioxide via the Calvin cycle. These organisms, which occur in soil, polluted waters and sewage, include e.g. *Bacillus schlegelii* and *Pseudomonas carboxydovorans*.

6.2   CARBON ASSIMILATION IN HETEROTROPHS

Collectively, the heterotrophs can assimilate a vast range of carbon sources; only a few examples are given below.

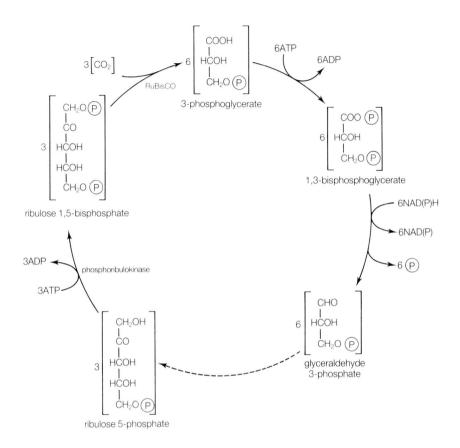

Fig. 6.1 Part of the Calvin cycle, showing the reaction in which carbon dioxide is 'fixed'. The purpose of the pathway is to enable carbon dioxide to be used for the synthesis of the complex carbon compounds which make up the cell itself. Essentially, carbon enters the cycle by the fixation of carbon dioxide, and carbon leaves the cycle when one or other of the intermediates is withdrawn for use in biosynthesis. For example, 3-phosphoglycerate is used for the synthesis of certain amino acids (such as glycine and serine) and also for the formation of pyruvate— itself a precursor of alanine, leucine and other amino acids; note that obligate autotrophs must synthesize all the amino acids necessary for protein synthesis. Continued operation of the Calvin cycle demands a continual supply of ribulose 1,5-bisphosphate; hence, the amount of carbon removed from the cycle must not exceed that put into the cycle. (The encircled 'P' represents phosphate.) RuBisCO is the enzyme ribulose 1,5-bisphosphate carboxylase–oxygenase. The many interconnecting pathways in the Calvin cycle have been omitted for clarity.

Many heterotrophs can use glucose, and its mode of assimilation will depend on the pathways and enzyme systems in any given organism. For example, many bacteria (including *E. coli*) can assimilate glucose via 'energy' pathways such as the EMP pathway (Fig. 5.3) and—subsequently—the TCA cycle (Fig. 5.10). In *Pseudomonas aeruginosa* glucose can be assimilated via extracytoplasmic oxidation (section 5.3.4). These pathways provide both energy and the compounds used as starting points in biosynthesis.

Some bacteria (e.g., species of *Acetobacter* and *Pseudomonas* can assimilate glucose via the *Entner–Doudoroff pathway* (Fig. 6.2); this yields NAD(P)H for biosynthesis as well as some useful precursor molecules. The *hexose monophosphate pathway* (HMP pathway) provides a source of ribulose 5-phosphate (Fig. 6.2)—an important precursor of nucleotides (Chapter 7) and of the amino acid histidine.

In *E. coli*, the sugar lactose is split into glucose and galactose. The glucose is metabolized by the EMP pathway (Fig. 5.3), and the galactose is metabolized by the Leloir pathway. In the *Leloir pathway*, galactose is 1-phosphorylated (by ATP), and the galactose 1-phosphate is converted, enzymatically, via glucose 1-phosphate to glucose 6-phosphate—which enters the EMP pathway.

Cellulose, a polymer of glucose, is used as a carbon source by a number of actinomycetes (bacteria found in soil and compost heaps etc.) and e.g. by certain bacteria in the *rumen* (the 'grass-digesting' part of the alimentary canal in cows and other ruminants). These bacteria produce enzymes which can degrade certain types of cellulose—outside the cell—into products which include glucose. Cellulolytic (i.e. cellulose-degrading) bacteria include species of *Cellulomonas* and e.g. *Clostridium thermocellum* and some strains of *Pseudomonas* and *Ruminococcus*.

6.3 SYNTHESIS, INTERCONVERSION AND POLYMERIZATION OF CARBON COMPOUNDS

6.3.1 Synthesis of carbon compounds

Synthesis of the vast range of molecules which form the structure of a living cell clearly demands an enormously complex (and rigorously controlled) network of chemical reactions. Moreover, as well as 'structural' molecules

---

Fig. 6.2 The Entner–Doudoroff pathway: a pathway which can be used for the assimilation of carbon in some heterotrophic bacteria. The hexose monophosphate pathway is similar as far as 6-phosphogluconate; it continues (as shown) via ribulose 5-phosphate. Glyceraldehyde 3-phosphate may be converted to pyruvic acid (as in the EMP pathway: Fig. 5.3) or (e.g. in pseudomonads) re-cycled, re-entering the pathway as 6-phosphogluconate.

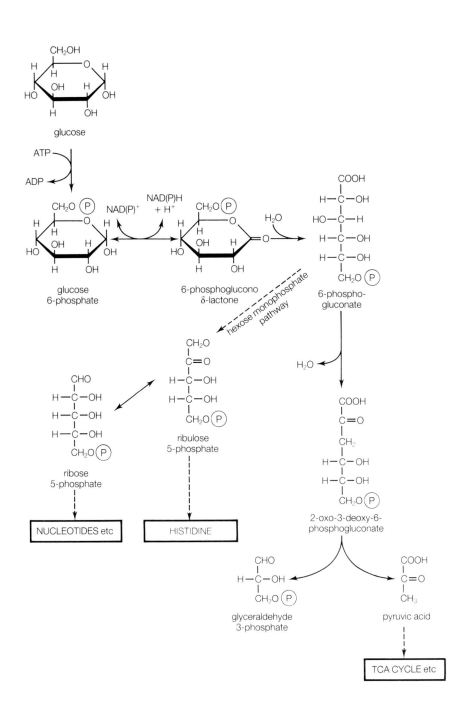

the cell needs energy with which to carry out the various anabolic reactions; carbon and energy metabolism are usually closely inter-linked (section 5.3.1). Here, we have space to look only briefly at some of the generalities of biosynthesis.

Many compounds in the initial assimilative pathways can be used more or less immediately as starting points for biosynthesis; from the cell's point of view this makes good sense: if lengthy metabolism were needed to make the carbon available, any early breakdown in the carbon pathway could be a serious problem for biosynthesis. Thus, for example, in the Calvin cycle (Fig. 6.1) the very first product of carbon dioxide fixation, 3-phosphoglycerate, can be used to synthesize the amino acids cysteine, glycine and serine (themselves components of proteins—section 7.6) as well as other compounds. Pyruvic acid (from the EMP and Entner-Doudoroff pathways) is a precursor of e.g. various amino acids, and ribose 5-phosphate (from the HMP pathway) is used e.g. to synthesize nucleotides (components of DNA and RNA—Chapter 7).

Intermediates in the TCA cycle can also be used for biosynthesis. For example, oxaloacetic and 2-oxoglutaric acids are precursors of a range of amino acids; acetyl-CoA can be used e.g. for fatty acid synthesis; and succinyl-CoA is used for the synthesis of porphyrins—components of e.g. chlorophylls and cytochromes.

Since intermediates withdrawn from the TCA cycle cannot be used to regenerate oxaloacetic acid (OAA), this compound must be generated in some other way if the cycle is to continue (Fig. 5.10). There are various reactions for achieving this—for example, under certain conditions, OAA may be formed directly by the carboxylation of pyruvic acid or of phosphoenolpyruvate; such 'replenishing' reactions are called *anaplerotic sequences*.

In general, the assimilative pathways in both autotrophs and heterotrophs yield a range of 3-, 4- and 5-carbon molecules which are useful precursors in biosynthesis. Each step in a biosynthetic reaction is normally enzyme-mediated, very few reactions occurring spontaneously. Reductive steps typically involve NAD(P)H or $FADH_2$, oxidative steps NAD or FAD, and phosphorylation usually involves ATP, GTP or phosphoenolpyruvate. Common coenzymes include coenzyme A (a carrier of acetyl and other acyl groups) and thiamine pyrophosphate (involved e.g. in various decarboxylation reactions).

Some examples of simple biosynthetic pathways are shown in Fig. 6.3.

6.3.2   Interconversion of carbon compounds

In many pathways, individual reactions—or even sequences of reactions—can go in either direction, depending on conditions. Moreover, intermediates (as well as end-products) can pass from one pathway to another, so that the flow of carbon into various products can be regulated according to the cell's requirements. This versatility is indicated by the fact that, in many cases, the great array of structural molecules can be synthesized from a single carbon source. Only a few brief examples of interconvertibility can be mentioned here.

The interconvertibility of carbon compounds is well illustrated by *gluconeogenesis*: the synthesis of glucose 6-phosphate from non-carbohydrate substrates such as acetate, glycerol or pyruvate. Essentially, this is achieved by conversion of the substrate (where necessary) to an intermediate in the EMP pathway (Fig. 5.3)—followed by reversal of that pathway; non-reversible reactions in the EMP pathway are by-passed by other enzymes. Bacteria capable of gluconeogenesis (e.g. *E. coli*) can, if necessary, use intermediates (as well as the end-product) in the reversed EMP pathway.

Where the hexose monophosphate and Entner–Doudoroff pathways occur in the same cell, the common intermediate 6-phosphogluconate (Fig. 6.2) may be metabolized by either pathway. Increased metabolism via the HMP pathway may occur, for example, during chromosome replication—DNA synthesis requiring increased production of ribose 5-phosphate, a component of both purine and pyrimidine nucleotides (section 7.2).

6.3.3 Polymerization of carbon compounds

Polymerization involves the chemical linkage of small molecules to form a large, often chain-like molecule called a *polymer*; a *homopolymer* is formed when all the small molecules are similar, and a *heteropolymer* results when the small molecules are not all alike. In bacteria, polymerization is involved in the synthesis of certain storage compounds and in the formation of various cell wall and capsular structures. Two (hetero)polymers of major importance—proteins and nucleic acids—are considered in the next chapter.

6.3.3.1 *Peptidoglycan synthesis*

In *E. coli*, N-acetylglucosamine and N-acetylmuramic acid (Fig. 2.7) are synthesized separately, in the cytoplasm, as their UDP (uridine 5'-diphosphate) derivatives—here abbreviated to UDP-GlcNAc and UDP-MurNAc, respectively. A chain of five amino acids is then added to UDP-MurNAc to form the so-called 'Park nucleotide'. The Park nucleotide is transferred to a long-chain lipophilic molecule (a *bactoprenol*) in the cytoplasmic membrane, and subsequently UDP-GlcNAc is added to form a disaccharide–pentapeptide subunit; during this stage the uridine moiety is lost by both sugars. Subunits are joined together, by *transglycosylation* reactions, to form part of a nascent backbone chain. Cross-links are then formed, by *transpeptidation*, between peptides on the nascent chain and those

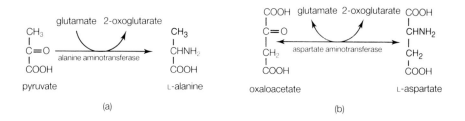

(a)   (b)

$$\begin{array}{c} CH_2O\,\text{(P)} \\ | \\ CHOH \\ | \\ COOH \end{array}$$

3-phosphoglycerate

NAD⁺ ⤵ phosphoglycerate
dehydrogenase

NADH + H⁺ ⤸

$$\begin{array}{c} CH_2O\,\text{(P)} \\ | \\ C{=}O \\ | \\ COOH \end{array}$$

3-phosphohydroxy-
pyruvate

glutamate ⤵ phosphoserine
aminotransferase

2-oxoglutarate ⤸

$$\begin{array}{c} CH_2O\,\text{(P)} \\ | \\ CHNH_2 \\ | \\ COOH \end{array}$$

3-phosphoserine

H₂O ⤵ phosphoserine
phosphatase

Pi ⤸

$$\begin{array}{c} CH_2OH \\ | \\ CHNH_2 \\ | \\ COOH \end{array}$$

L-serine

(c)

in pre-existing peptidoglycan. Transglycosylation and transpeptidation involve the penicillin-binding proteins (section 2.2.8).

6.3.3.2 Synthesis of poly-β-hydroxybutyrate

When non-carbon nutrients become scarce, many bacteria form intracellular granules of poly-β-hydroxybutyrate (PHB)—degrading these granules when conditions return to normal; PHB thus acts as a reserve of carbon and/or energy. Typically, PHB (Fig. 2.3) is synthesized from acetyl-CoA via acetoacetyl-CoA and β-hydroxybutyryl-CoA; the polymerizing enzyme, PHB synthetase, occurs bound to the thin membrane which normally covers the granules.

Other polymers of β-hydroxyalkanoates occur in some bacteria; thus, e.g. *Bacillus megaterium* may form small amounts of β-hydroxyheptanoate with the PHB.

6.4 METHYLOTROPHY

Methylotrophy is the obligate or facultative use of so-called 'C$_1$ compounds' as the sole source of carbon and energy. A C$_1$ compound is a carbon compound in which the molecule has no carbon–carbon bonds and which is in a more reduced state than carbon dioxide; C$_1$ compounds include carbon monoxide, methane, methanol and formaldehyde. *Methylotrophs* (organisms capable of methylotrophy) include e.g. species of *Hyphomicrobium* and *Methylococcus*, and the carboxydobacteria (section 6.1.3).

C$_1$ substrates are oxidized to products which include formaldehyde and/or carbon dioxide. Carbon is assimilated as formaldehyde (via specialized pathways) and/or as carbon dioxide (either via the Calvin cycle or an alternative pathway).

One obligate methylotroph, *Methylophilus methylotrophus*, used to be grown commercially for use as an animal feed; thousands of tons of the product ('Pruteen') were made each year—until protein prices fell and the process was discontinued.

Fig. 6.3 Some simple biosynthetic pathways: synthesis of the amino acids L-alanine (a), L-aspartate (b) and L-serine (c). (*Note*. Biochemists often give the *formula* of an organic acid in the un-ionized form but *name* the compound as though it were the salt—e.g. CH$_3$.CO.COOH = pyruvate.) In these particular examples, the reaction with glutamate introduces nitrogen into the (non-nitrogenous) precursor molecule; glutamine carries out this function in the synthesis of L-tryptophan. Note that each reaction is catalysed by a specific enzyme. An encircled 'P' represents a phosphate group; 'Pi' is inorganic phosphate.

7 Molecular biology I: genes and gene expression

7.1 CHROMOSOMES AND PLASMIDS

The chromosome (section 2.2.1) consists mainly of a polymer called *deoxyribonucleic acid* (DNA). DNA and a related polymer, ribonucleic acid (RNA), belong to a category of molecules (*nucleic acids*) which can carry information; in these molecules information is carried in the *sequence* with which subunits of the polymer are fitted together (see later). Chromosomal DNA carries all the information needed to specify both the structure and behaviour of a bacterium.

DNA dictates the life of a cell e.g. by encoding all the enzymes (thus controlling structure and metabolism) and by encoding various RNA molecules (involved in protein synthesis and certain control functions). Information carried by DNA regulates and co-ordinates growth and differentiation. Moreover, DNA controls its own replication and is also a self-monitoring system: there are various mechanisms for detecting and repairing damaged or altered DNA. All of this information is passed to daughter cells when the chromosome replicates and the parent cell divides (section 3.2.1). During replication the DNA is normally copied very accurately so that the characteristics of the species remain stable from one generation to the next; the tendency of daughter cells to inherit parental characteristics—heredity—is the main focus of *genetics*.

Clearly, nucleic acids determine the life of a cell as well as the process of heredity; both of these roles are studied in *molecular biology*.

Many bacteria contain one or more *plasmids*; a plasmid is an 'extra' piece of DNA which is much smaller than the chromosome and which can replicate independently. Plasmids can encode various functions. Some encode enzymes which inactivate particular antibiotics; such *R plasmids* (resistance plasmids) usually make the host cell resistant to the relevant antibiotics. Some plasmids encode structural elements—e.g. gas vacuoles (section 2.2.5) in certain strains of *Halobacterium*. The 'Cit' plasmid encodes a transport system (section 5.4) for the uptake of citrate; strains of *E. coli* which contain this plasmid can use citrate as the sole source of carbon and energy— something which common or 'wild-type' strains cannot do. (Other plasmid-mediated functions are mentioned elsewhere in the book.) However,

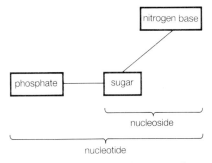

Fig. 7.1 A generalized nucleotide. Note the difference between a nucleotide and a nucleoside.

although plasmids often encode useful functions they are typically not indispensable to their host cells.

Some (not all) plasmids encode the means to transfer themselves from one cell to another by,the process of *conjugation* (section 8.4.2).

Prophages (section 9.2) which do not integrate with the host's chromosome, and which can replicate, are also called 'plasmids'.

7.2 NUCLEIC ACIDS: STRUCTURE

A nucleic acid is a polymer made of subunits called *nucleotides*; each nucleotide has three parts: (i) a sugar molecule, (ii) a nitrogen-containing base, and (iii) a phosphate group (Fig. 7.1). Nucleotides containing the sugar D-ribose are *ribo*nucleotides—which form *ribo*nucleic acid (RNA); those containing 2'-deoxy-D-ribose are *deoxyribo*nucleotides—which form *deoxyribo*nucleic acid (DNA)(Fig. 7.2). The nitrogen base in any given nucleotide is either a substituted *purine* or a substituted *pyrimidine* (Fig. 7.3); a ribonucleotide may

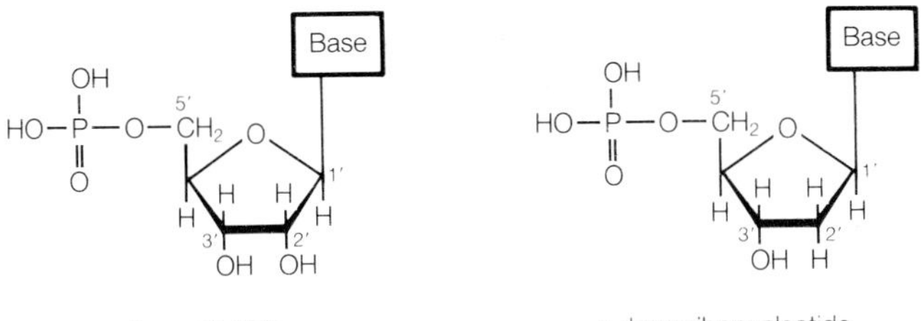

a ribonucleotide   a deoxyribonucleotide

Fig. 7.2 Nucleotides: a ribonucleotide monophosphate and a deoxyribonucleotide monophosphate; the molecules differ only in their sugar (ribose) residues: the deoxyribonucleotide lacks oxygen at the 2'-position. The 'base' can be adenine, guanine or cytosine in either molecule, but uracil occurs only in ribonucleotides, while thymine is found only in deoxyribonucleotides.

Table 7.1   Nucleosides and nucleotides found in RNA and DNA

Sugar	Nitrogen base	Nucleoside (base + sugar)	Nucleotide[1] (base + sugar + phosphate)
Ribose	Adenine	Adenosine	Adenosine 5'-monophosphate (AMP) (or adenylic acid)
Ribose	Guanine	Guanosine	Guanosine 5'-monophosphate (GMP) (or guanylic acid)
Ribose	Cytosine	Cytidine	Cytidine 5'-monophosphate (CMP) (or cytidylic acid)
Ribose	Uracil	Uridine	Uridine 5'-monophosphate (UMP) ((or uridylic acid)
Deoxyribose	Adenine	Deoxyadenosine	Deoxyadenosine 5'-monophosphate (dAMP) (or deoxyadenylic acid)
Deoxyribose	Guanine	Deoxyguanosine	Deoxyguanosine 5'-monophosphate (dGMP) (or deoxyguanylic acid)
Deoxyribose	Cytosine	Deoxycytidine	(Deoxycytidine 5'-monophosphate (dCMP) (or deoxycytidylic acid)
Deoxyribose	Thymine	Deoxythymidine	Deoxythymidine 5'-monophosphate (dTMP) (or deoxythymidylic acid)

[1]Only the monophosphate is shown in each case. Diphosphates and triphosphates are named in a similar fashion, e.g. adenosine 5'-diphosphate (ADP) and adenosine 5'-triphosphate (ATP).

contain adenine, guanine, cytosine or uracil, while a deoxyribonucleotide may contain adenine, guanine, cytosine or thymine. The names of the various nucleotides, and their corresponding nucleosides, are given in Table 7.1.

7.2.1   Deoxyribonucleic acid (DNA)

In DNA the nucleotides form an unbranched chain (a *strand*) in which sugar–base and phosphate residues are alternate links (Fig. 7.4). Note that a strand has *polarity*: there is a 5'-end and a 3'-end. DNA commonly consists of two strands which are held together by hydrogen bonding between their nitrogen bases; this double-stranded structure is called a DNA *duplex* (Fig. 7.5). Note that the two strands in a duplex are *antiparallel*, i.e. when read left to right (Fig. 7.5), one strand is 5'-to-3' while the other is 3'-to-5'.

Hydrogen bonding between the nitrogen bases is quite specific: adenine pairs with thymine, and guanine pairs with cytosine (Fig. 7.6); this specificity in *base-pairing* is referred to by saying that each of the two bases in a *base-pair* is *complementary* to its partner, and that (therefore) each strand in a DNA duplex is complementary to the other strand.

The ladder-like DNA duplex is itself twisted into a *helix* (Fig. 7.7)—each turn of which occupies about 10 base-pairs of distance along the duplex; this is the 'double helix' worked out by Watson and Crick in the 1950s.

In the chromosome, and in most (not all) plasmids, the helical DNA duplex

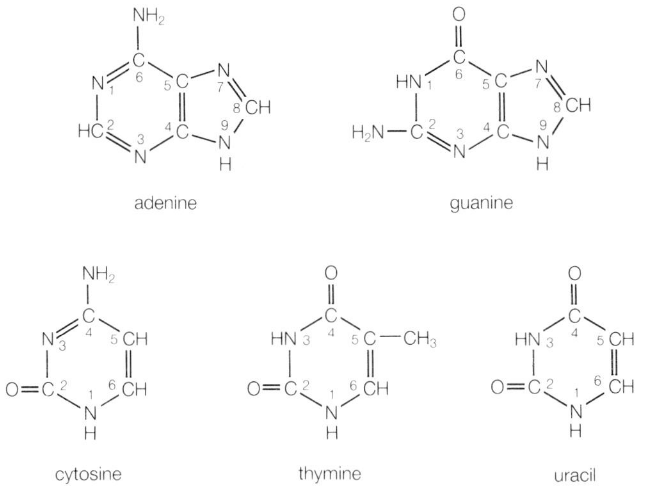

Fig. 7.3 Nitrogen bases found in DNA and RNA. Adenine and guanine are substituted *purines*; the others are substituted *pyrimidines*. Thymine occurs only in DNA, uracil only in RNA; adenine, guanine and cytosine occur in both DNA and RNA. Within a nucleotide (Fig. 7.2) a purine is linked via its 9-position to the sugar molecule, while a pyrimidine is linked via its 1-position.

forms a closed loop, i.e. there are no free 5' and 3' ends; this is double-stranded, covalently-closed circular DNA: ds cccDNA. Such DNA can exist in various states. In the *relaxed* state, a loop of DNA can theoretically lie in a plane—like a rubber band on a table. Suppose, however, that the loop is cut, and that one end is twisted (the other held still) before the ends are re-joined; this will either increase or decrease the 'pitch' of the helix (forming an 'overwound' or 'underwound' helix, respectively) depending on the direction of twisting. Such a molecule is under strain: there is a tendency to restore the pitch to that of a relaxed molecule. To relieve the strain, the molecule contorts: the axis of the helix becomes helical, i.e. the helix itself coils up to form a helix! A molecule in this state is said to be *supercoiled*.

Naturally occurring supercoiled DNA is generally 'underwound' (i.e.

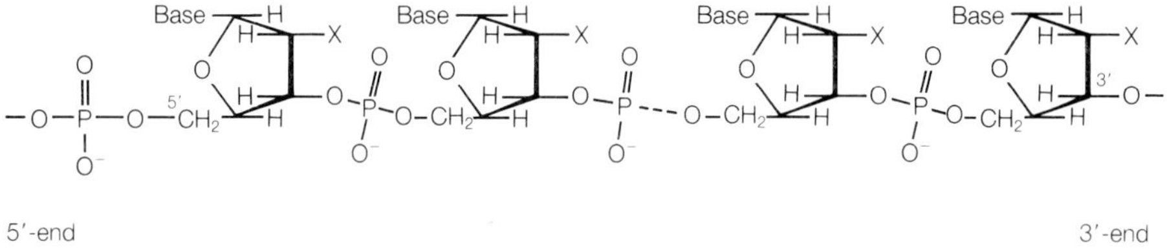

5'-end

3'-end

Fig. 7.4 The structure of a single strand of nucleic acid; note the polarity of the molecule. 'X' is a hydrogen atom (H) in DNA but a hydroxyl group (OH) in RNA. The sugar residues are linked together by *phosphodiester bonds*.

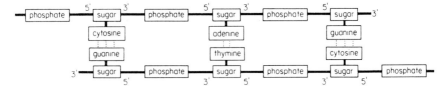

Fig. 7.5 A DNA duplex (diagrammatic). The dotted lines represent hydrogen bonds. Note that the strands are *antiparallel* (section 7.2.1).

'negatively supercoiled'). The degree of supercoiling is controlled by enzymes called *topoisomerases*.

7.2.2 Ribonucleic acid (RNA)

RNA is a linear polymer of ribonucleotides (Fig. 7.4); although the ribonucleotides generally contain adenine, guanine, cytosine or uracil, modified bases occur in some RNA molecules. Bacterial RNA (unlike DNA) is typically single-stranded; however, double-stranded regions are formed in some cases by base-pairing between complementary sections within the same molecule.

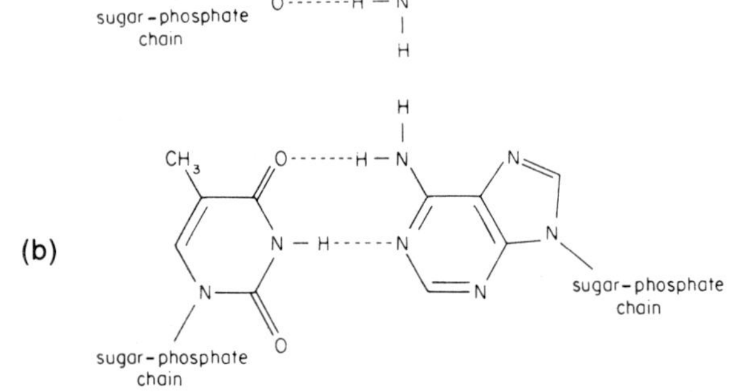

(a)

(b)

Fig. 7.6 Base-pairing between nucleotides. (a) Cytosine (left) pairing with guanine (right). (b) Thymine (left) pairing with adenine (right). (*Note.* In RNA synthesis (section 7.5) adenine pairs with uracil.) The dotted lines represent hydrogen bonds.

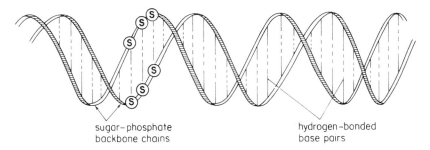

sugar–phosphate
backbone chains

hydrogen–bonded
base pairs

Fig. 7.7 The DNA 'double helix' (diagrammatic). An encircled 'S' represents a sugar residue within the backbone chain; sugar residues are linked by phosphodiester bonds (Fig. 7.4). The (planar) nitrogen bases are stacked roughly perpendicular to the axis of the helix; chemical groups on these bases protrude into the grooves of the helical molecule—thus allowing e.g. recognition by enzymes.

7.3   DNA REPLICATION

As mentioned earlier, the *sequence* of bases in chromosomal DNA is a coded message specifying the cell's structure and behaviour. Before cell division occurs, the DNA must be duplicated precisely so that each daughter cell will receive an exact copy (replica) of the molecule. DNA synthesis (*replication*) is a complex process; the following is an outline, only, of the process in the *E. coli* chromosome.

Replication starts at a specific site in the DNA molecule, the *origin*; various initiation factors and enzymes are needed—e.g. *topoisomerase I* (which relaxes negatively supercoiled DNA), *gyrase* (which can catalyse negative supercoiling), and an enzyme which polymerizes ribonucleotides, *RNA polymerase*. Locally, in the region of the origin, the two strands of the duplex separate. Newly exposed bases in *one* of the strands then pair, sequentially, with the complementary bases of free, individual *ribo*nucleotides; the ribonucleotides are polymerized (in the 5'-to-3' direction) by RNA polymerase to form a short strand of RNA, a *primer*. The primer thus forms one strand of a short hybrid (RNA/DNA) duplex (Fig. 7.8, top). As the DNA duplex continues to open, *deoxy*ribonucleotides base-pair with newly exposed bases, being sequentially added to the primer in the 5'-to-3' direction; the deoxyribonucleotides are polymerized by an enzyme called *DNA polymerase III* to form the start of the so-called 'leading' strand of DNA (Fig. 7.8). Note that the leading strand—together with one strand of the original duplex—forms the start of a new DNA duplex. Synthesis of the new DNA strand exploits the specificity of base-pairing (section 7.2.1), and the 'original' strand (which determines bases in the new strand) is called the *template* strand. An RNA primer is required to start the process because DNA polymerases can only extend an *existing* strand, i.e. they cannot initiate a strand.

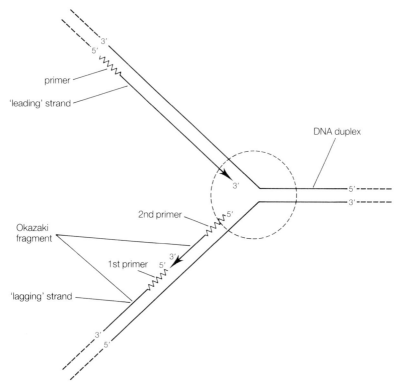

Fig. 7.8 DNA replication (Cairns-type, diagrammatic). In a circular DNA duplex the molecule has split, locally, into its two component strands. The diagram shows only one half of the split region; this half continues to split in a left-to-right direction. On one exposed DNA strand (top) an RNA primer is synthesized (see text); the subsequent addition of deoxyribonucleotides forms the 'leading' strand of DNA—which is synthesized continuously in the 5'-to-3' direction. On the other DNA strand (bottom) the first primer is extended by synthesis of a DNA fragment (an *Okazaki fragment*)—again in the 5'-to-3' direction (because synthesis cannot occur in the 3'-to-5' direction); only when the duplex has opened out further is a second primer formed and the next Okazaki fragment synthesized. (All primers are subsequently replaced with DNA.) The encircled region (*replication fork*) moves to the right as replication continues, while the other replication fork (not shown) moves to the left.

What of the other strand of the original DNA duplex? That, too, acts as a 'template'. However, that particular strand (Fig. 7.8) has a different *polarity* (section 7.2.1), and DNA cannot be synthesized in the 3'-to-5' direction. Because of this, the other new strand of DNA (the 'lagging' strand) is synthesized as a series of short fragments (*Okazaki fragments*), as follows. When strand-separation in the DNA duplex has reached a certain extent, the enzyme *DNA primase* synthesizes an RNA primer near the replication fork; the primer is extended by deoxyribonucleotides, in the 5'-to-3' direction, to form the first fragment of the new 'lagging' strand of DNA (Fig. 7.8). On further strand-separation the second primer is formed and the second fragment of DNA is synthesized—and so on.

A replication fork (Fig. 7.8) is also formed on the other side of the origin; hence, DNA replication proceeds in *both* directions from the origin.

Before completion of replication all primers are replaced with DNA.

Nucleotides used for the synthesis of the new DNA strands, and the primers, are nucleotide *tri*phosphates; the two terminal phosphate groups are cleaved to provide energy for the polymerization reaction.

Replication yields two supercoiled DNA duplexes. Since each duplex consists of one strand of the original ('parent') duplex and one new ('daughter') strand, replication is said to be *semi-conservative*.

7.3.1 Replication of plasmid DNA

A plasmid controls its own replication; however, the cell's biosynthetic machinery is needed to synthesize any plasmid-encoded protein etc. involved in replication.

The frequency of replication depends on the rate of *initiation*, and this is controlled in different ways by different plasmids. In plasmid ColE1, for example, control involves the synthesis of two, free (non-duplexed) strands of plasmid-encoded RNA, one of which (*RNA II*) can bind at the plasmid's origin and act as a primer, thus permitting replication; the other strand (*RNA I*) can interfere by base-pairing with RNA II, and the outcome of RNA I-RNA II interaction determines the occurrence of replication. Thus, in ColE1, initiation is controlled at the primer level.

Replication of plasmid R1162 depends on the binding of a plasmid-encoded initiator, the *RepIB protein*, to certain 'direct repeat' nucleotide sequences (*iterons*) in the plasmid's origin of replication; it has been shown that RepIB–iteron interaction brings about localized 'melting' (i.e., strand separation) in the DNA duplex [Kim & Meyer (1991) JB *173* 5539–5545]. Here, initiation seems to be controlled at the level of strand separation at the replicative origin.

In the F plasmid, control of replication involves the plasmid-encoded *E protein*.

To understand replication control in plasmids it is necessary to know whether control molecules (such as the RepIB and E proteins) are synthesized continually or periodically, how their synthesis/activity is regulated, and whether other factors/molecules are involved.

Once initiated, replication may proceed as in the *E. coli* chromosome (*Cairns-type* DNA replication—see above). This occurs e.g. in ColE1, although in this plasmid replication proceeds *uni*directionally from the origin. Replication is bidirectional in the F plasmid.

In a host cell, the number of copies of a given plasmid, per chromosome, is called the *copy number*. Copy number depends on the plasmid's replication control system, on the bacterial strain and on growth conditions. Some plasmids (e.g. the F and R6 plasmids) have a copy number of 1–2; in *multicopy plasmids* (e.g. ColE1, R6K) it may be e.g. 10–30.

7.3.1.1 Stringent and relaxed control

Some plasmids (e.g. the F plasmid) fail to replicate if their host cells are treated with an antibiotic (e.g. chloramphenicol) which inhibits protein synthesis (section 7.6); the replication of such plasmids is said to be under *stringent control*. Other plasmids (e.g. ColE1) continue to replicate under these conditions—and can achieve higher-than-normal copy numbers; the replication of such plasmids is said to be under *relaxed control*.

7.4 DNA MODIFICATION AND RESTRICTION

In many (perhaps all) bacteria, each newly replicated strand of DNA undergoes chemical 'modification': methylation of certain bases in *specific sequences of nucleotides*; in these sequences the methyl group (CH_3–) may occur at the N-6 position of adenine and/or the C-5 position of cytosine (Fig. 7.3). Different sequences of nucleotides are methylated in different strains of bacteria. Methylation protects DNA from the cell's own *restriction endonucleases*: enzymes which cleave ('restrict') any DNA lacking methylation in appropriate sequences of nucleotides. A duplex is not cleaved if at least one of the strands has been methylated; normally, the (methylated) template strand protects the new daughter strand until the latter has been methylated. Restriction acts primarily against 'foreign' DNA which has entered the cell. [Review: Wilson & Murray (1991) ARG 25 585–627.]

Each different restriction endonuclease (RE) recognizes a specific sequence of nucleotides (*recognition sequence*). REs are classified as types I, II and III according e.g. to the sites at which they cleave an (unmethylated) duplex. A type II RE cleaves the duplex within, or very close to, its recognition sequence; type II REs, which need Mg^{2+} but not ATP for activity, are widely used in genetic engineering.

REs are named as in the following example: *Eco*RI (RE from *E. coli* strain R, 'I' indicating a particular RE from strain R); *Eco*RI recognizes the sequence GAATTC (written 5'-to-3') and cleaves the duplex between G and A. (See also Table 8.1.)

7.5 RNA SYNTHESIS

In bacterial RNA synthesis, individual *ribo*nucleotides base-pair with exposed bases on a DNA (not RNA) template strand and are polymerized by an RNA polymerase; note that, in this case, adenine pairs with uracil, not with thymine. As in DNA synthesis, nucleotide *tri*phosphates base-pair with the template strand—the two terminal phosphate groups being cleaved in a reaction which provides energy for polymerization. The synthesis of an RNA strand on a DNA template is called *transcription*; the RNA strand itself is called a *transcript*.

Transcription is initiated when an RNA polymerase, bound to another protein called a *sigma factor*, binds at a specific sequence of nucleotides (a *promoter*) in a DNA duplex. A cell has a number of different sigma factors, each enabling the RNA polymerase to recognize a particular type of promoter. Locally, the strands of the DNA duplex separate, and the first ribonucleotide base-pairs with a complementary base on one of the DNA strands at a site called the *start point*; the polymer grows as ribonucleotides are added, sequentially, in the 5'-to-3' direction. After a few ribonucleotides have been polymerized, the sigma factor is released.

Elongation involves progressive unwinding of the DNA duplex and the sequential addition of ribonucleotides. As the process continues, the RNA strand peels away from the template strand and the DNA duplex re-forms.

Transcription stops at a specific sequence of nucleotides (a termination signal, or *terminator*) in the template strand. In so-called *rho-independent* termination, parts of the transcript of the terminator region base-pair with one another, and the resulting structure may cause the release of the transcript and/or polymerase. In *rho-dependent* termination, a protein, the *rho-factor*, may stop transcription after recognizing a particular site on the transcript.

The initial product of transcription (the *primary transcript*) may require processing to give the final, mature RNA transcript.

The synthesis of a variety of RNA molecules is needed for protein synthesis (section 7.6), and strands of RNA are also needed, as primers, in DNA synthesis (section 7.3). Additionally, some RNA molecules are synthesized for specific control functions; these *antisense RNA* molecules can inhibit the activity of other nucleic acids by base-pairing to complementary sequences in them.

7.6   PROTEIN SYNTHESIS

All the cell's proteins are encoded by DNA; by studying protein synthesis we can see *how* DNA carries this information—and how the genetic code works.

A protein consists of one or more *polypeptides*, a polypeptide being a chain of amino acids covalently linked by peptide bonds (–CO.NH–). Each polypeptide is folded into a three-dimensional structure; this structure is stabilized mainly by hydrogen bonds or disulphide bonds formed between amino acids in different parts of the chain. The specific three-dimensional structure of a given polypeptide—essential for biological activity—is determined by the nature, number and sequence of its amino acids.

For any given polypeptide, the nature, number and sequence of amino acids are dictated by a particular sequence of bases in a DNA strand; this sequence of bases conforms to one definition of a *gene*.

How does a gene bring about the synthesis of a polypeptide? That is, how is the gene *expressed*? Unlike nucleotides, amino acids cannot simply 'line up' (undergo polymerization) on a DNA template strand: apart from other considerations, amino acids cannot base-pair with nitrogen bases. In fact, protein synthesis involves several stages. First, the gene is *transcribed* (section 7.5); the (RNA) transcript of the gene, which carries the 'message' from DNA, is called *messenger RNA* (mRNA). Now, along the length of the mRNA molecule, groups of three consecutive bases each encode a particular amino acid; each of these three-base groups is called a *codon*. Thus, e.g. the codon UCA (uracil–cytosine–adenine) encodes the amino acid serine (Table 7.2). Hence, the sequence of codons in mRNA encodes the sequence of amino acids in a polypeptide. For the synthesis of a polypeptide, each amino acid must first bind to an 'adaptor' molecule which is specific for that particular amino acid and for its codon; these 'adaptor' molecules are small molecules of RNA—*transfer RNA* (tRNA)—and each binds to its specific amino acid to form an *aminoacyl-tRNA*. (Such binding requires ATP.) As well as a binding site for the amino acid, a given tRNA molecule also contains a sequence of three bases (an *anticodon*) which is complementary to the codon of its particular amino acid; hence, a given tRNA molecule can bind its particular amino acid and can transfer it to the specific codon in mRNA through codon–anticodon base-pairing.

The synthesis of a polypeptide on mRNA (the *translation* process) takes place on a ribosome (section 2.2.3); a simplified version of this process is shown diagrammatically, and explained, in Fig. 7.9.

As is indicated in Fig. 7.9, the *initiator codon* (i.e. the first codon to be translated) is commonly AUG; when acting as an initiator codon, AUG encodes the modified amino acid N-formylmethionine in most or all eubacteria (including *E. coli*). (The formyl group is later removed from the polypeptide.)

Alignment of AUG with the 'P' site is generally promoted by a sequence of nucleotides (the *Shine–Dalgarno sequence*) 'upstream' of AUG on the mRNA molecule (i.e. to the left of AUG in Fig. 7.9); this sequence base-pairs with part of a 16S rRNA molecule in the ribosomal 30S subunit.

Other codons are usually translated according to the *genetic code* (Table 7.2). UAA, UAG and UGA (*ochre, amber* and *opal* codons, respectively) are 'stop' codons which signal termination of polypeptide synthesis; they are also called *nonsense codons*.

As one ribosome translocates along the mRNA, another can fill the

Table 7.2   The 'universal' genetic code: amino acids and 'stop' signals encoded by particular codons[1-5]

First base (5' end)	Second base				Third base (3' end)
	U	C	A	G	
U	Phe	Ser	Tyr	Cys	U
	Phe	Ser	Tyr	Cys	C
	Leu	Ser	*ochre*	*opal*	A
	Leu	Ser	*amber*	Trp	G
C	Leu	Pro	His	Arg	U
	Leu	Pro	His	Arg	C
	Leu	Pro	Gln	Arg	A
	Leu	Pro	Gln	Arg	G
A	Ile	Thr	Asn	Ser	U
	Ile	Thr	Asn	Ser	C
	Ile	Thr	Lys	Arg	A
	Met	Thr	Lys	Arg	G
G	Val	Ala	Asp	Gly	U
	Val	Ala	Asp	Gly	C
	Val	Ala	Glu	Gly	A
	Val	Ala	Glu	Gly	G

[1]A = adenine; C = cytosine; G = guanine; U = uracil.
[2]Standard abbreviations are used for the amino acids: Gln = glutamine, Ile = isoleucine etc.
[3]As with other nucleotide sequences, codons are conventionally written in the 5'-to-3' direction; as anticodons also are written in this way, the *first* base in a codon pairs with the *third* base in its anticodon.
[4]The codons UAA (*ochre*), UGA (*opal*) and UAG (*amber*) are normally 'stop' signals (see text).
[5]Although called 'universal', some exceptions to the code have been reported. For example in *E. coli* the codon UGA encodes seleno-cysteine; the tRNA which base-pairs with UGA binds serine, but the serine is converted to selenocysteine before incorporation in the polypeptide (*co-translation*) [Leinfelder *et al.* (1988) Nature *331* 723–725]. A recent review deals with the evolution of the genetic code [Osawa *et al.* (1992) MR *56* 229–264].

vacated initiation site and start translation of another molecule of the polypeptide; thus, a given mRNA molecule can carry a number of ribosomes along its length—forming a *polyribosome (polysome)* (Fig. 7.10).

In many cases, a protein destined to form part of a membrane, or to pass through a membrane, is synthesized with a special N-terminal sequence of amino acids (a *signal sequence*); this may help passage into, or through, the hydrophobic region of the membrane, and it may be excised once the protein is in place (an example of *post-translational modification*).

7.6.1   The fate of mRNA

After use, *bacterial* mRNA is typically short-lived, being rapidly degraded to

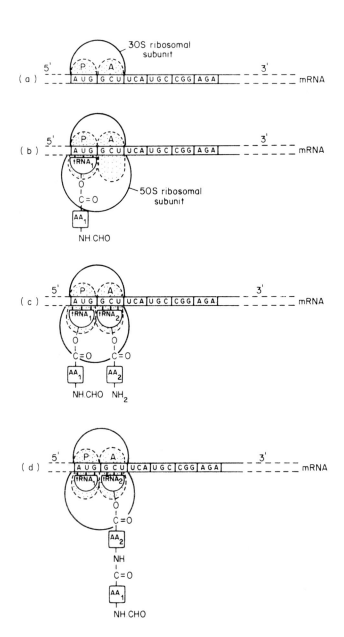

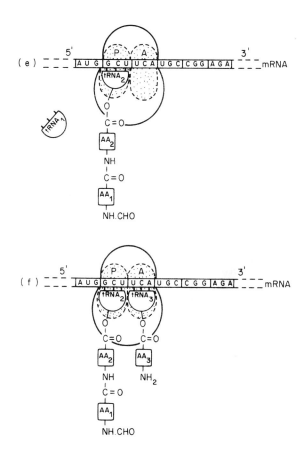

Fig. 7.9   Protein synthesis in bacteria (simplified, diagrammatic). (a) The 30S ribosomal subunit binds to mRNA with the *initiator codon* (AUG) at the 'P' site. (b) The first amino acid (AA$_1$), on its tRNA (tRNA$_1$), occupies the 'P' site, i.e. the anticodon of tRNA$_1$ base-pairs with codon AUG. The 50S ribosomal subunit binds, completing the ribosome. (c) The second aminoacyl-tRNA occupies the 'A' site, i.e. the anticodon of tRNA$_2$ base-pairs with codon GCU. (d) A peptide bond is formed between the carboxyl group of AA$_1$ and the α-amino group of AA$_2$; this is called *transpeptidation*. (Note that AA$_1$ will form the 'amino end' or 'N-terminal' of the polypeptide chain.) (e) The ribosome moves along the mRNA, by one codon, in the 5'-to-3' direction (*translocation*), and tRNA$_1$ is released; as a result, the dipeptidyl-tRNA now occupies the 'P' site, and there is a vacant 'A' (= 'acceptor') site opposite the third codon. (f) A third aminoacyl-tRNA binds at the 'A' site. Steps (d)–(f) are repeated for each codon in turn. When a 'stop' codon (e.g. UAA) is reached, a protein *release factor* hydrolyses the ester bond between the last tRNA and the polypeptide chain—thus releasing the completed chain.

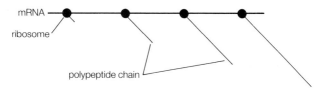

Fig. 7.10   A polysome (diagrammatic). As a given ribosome travels further along the mRNA molecule, its associated polypeptide chain increases in length.

re-cyclable components; this is essential: the cytoplasm would otherwise quickly fill up with 'used' mRNA molecules. Degradation involves *endonuclease(s)* (enzymes which cleave phosphodiester bonds in non-terminal parts of a nucleic acid strand) and *exonuclease(s)* (which sequentially cleave nucleotides from the 3' end). In *E. coli*, the exonucleases RNase II and polynucleotide phosphorylase jointly degrade mRNA to pieces of about 10 nucleotides in length—but no smaller; how, then, is mRNA re-cycled? Apparently, a previously unknown enzyme—RNase*—degrades these pieces to single nucleotides, thus completing the scheme for mRNA degradation [Cannistraro & Kennell (1991) JB *173* 4653-4659].

7.7   DNA MONITORING AND REPAIR

Abnormal DNA can result e.g. from the insertion of abnormal nucleotides during replication; it may be recognized and repaired immediately through *proof-reading*: DNA polymerase III can cleave a 'wrong' nucleotide from the 3' end of a growing strand, allowing replacement with a normal one.

Errors which escape proof-reading may be corrected later by the *mismatch repair* system. In *E. coli*, the MutS protein can recognize a single mismatched base-pair; the new (daughter) strand, identified by its transient under-methylation (section 7.4), is 'nicked' either side of the error, and an enzyme, apparently helicase II, removes the nicked section—which may be 1000 or more nucleotides in length. A DNA polymerase closes the gap by synthesizing on the parent (template) strand, and the junction between new and old strands is sealed by a ligase.

Mismatch repair can also correct mis-match in a 'heteroduplex' (section 8.2.1); if the latter is fully methylated, then *either* strand can apparently be corrected (i.e. made complementary to the other strand).

7.7.1   Excision repair

This process can repair damage affecting only *one* strand of dsDNA. In e.g. *E. coli*, an ATP-dependent enzyme, UvrABC endonuclease (= 'ABC excinuclease'), recognizes damaged/distorted DNA and cleaves a phospho-diester bond (Fig. 7.4) on each side of the damaged section (the 'incision' stage); the damaged section is removed (apparently by helicase II), and the gap is closed and sealed as in mismatch repair, above. In *E. coli*, only about 10 nucleotides in and around the damaged site are involved—hence the name *short patch repair*. [Nucleotide excision repair in *E. coli*: van Houten (1990) MR 54, 18–51.]

7.8   REGULATION OF GENE EXPRESSION

A cell does not express all of its genes all of the time. For example, if a particular substrate were *not* available the cell would be wasting energy if it synthesized those proteins (e.g. enzymes) needed to metabolize that substrate. In fact, many genes can be 'switched on' (*induced*) or 'switched off' (*repressed*) according to conditions. Repression commonly involves prevention of synthesis of the gene's mRNA (section 7.6); that is, gene expression is generally controlled 'at the level of transcription'.

7.8.1   Operons

In many cases a *sequence* of genes is transcribed—as a single unit—from a single promoter; a sequence of genes which is subject to co-ordinated expression in this way is said to form an *operon*. Genes in a given operon often encode functionally related products—e.g. the enzymes for a particular metabolic pathway. Various mechanisms are involved in controlling the expression of operons.

7.8.1.1   Operons under promoter control

In these operons, control involves a *regulator protein* which may be formed more or less continually. Operons which are expressed unless 'switched off' by the regulator protein are said to be under *negative control*; those which are not expressed unless 'switched on' by the regulator protein are under *positive control*.

The *lac* operon in *Escherichia coli* (Fig. 7.11) encodes proteins which promote the uptake and metabolism of β-galactosides such as lactose (a disaccharide composed of glucose and galactose residues); the presence of *allolactose* can 'induce' the *lac* operon (see Fig. 7.11). The *lacY* gene encodes 'β-galactoside permease' which promotes lactose uptake; *lacZ* encodes the enzyme β-galactosidase which can split lactose into glucose and galactose—but which also converts a small amount of lactose to allolactose. The *lacA* product (thiogalactoside transacetylase) appears not to be needed for lactose metabolism. The *lac* operon, which is under negative control, is explained in Fig. 7.11.

The *ara* operon in *Escherichia coli* (concerned with metabolism of the sugar

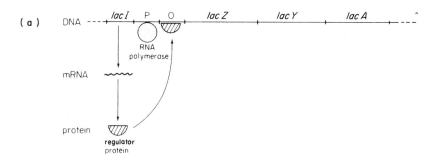

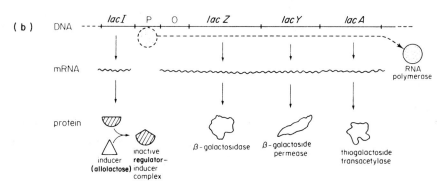

Fig. 7.11   The *lac* operon in *Escherichia coli*. (a) In the absence of an inducer, the regulator protein (product of the *lacI* gene) binds to the operator, O, and minimizes transcription of genes *lacZ*, *lacY* and *lacA* by the RNA polymerase (seen here at the promoter, P); there is some 'leakage': a few molecules of e.g. the *lacZ* product (β-galactosidase) are formed under these conditions. (b) In the presence of lactose (taken up e.g. by proton—lactose symport: section 5.4), the enzyme β-galactosidase converts some of the lactose to *allolactose* (section 7.8.1.1); allolactose acts as an inducer of the *lac* operon by binding to—and thereby inactivating—the regulator protein. Once the regulator protein has been inactivated, the three genes can be transcribed and translated. (For each of the three genes, the single mRNA transcript has an initiator codon and a stop codon—see section 7.6.)

When the lactose has been used up, transcription of the *lac* operon is no longer needed; the inducer (allolactose) is no longer formed, and the (now active) regulator protein 'switches off' the operon. The mRNA is rapidly degraded (section 7.6.1).

The *lac* operon is subject to catabolite repression (section 7.8.2.1).

arabinose) contains more than one operon; however, there is only one regulator protein: the AraC protein (encoded by the *araC* gene). In the absence of arabinose, AraC acts as a repressor (i.e. *negative* control, as in the *lac* operon). However, when arabinose is present it converts AraC to an *activator* which binds to an initiator and starts transcription; the (modified) regulator protein thus 'switches on' the system—an example of *positive* promoter control.

7.8.1.2   Operons under attenuator control

These operons are concerned e.g. with the synthesis of amino acids. In attenuator control, the initial sequence of nucleotides in the transcript (the *leader* sequence) encodes a small peptide (*leader peptide*) which is rich in the particular amino acid whose synthesis is governed by that operon; the leader sequence also encodes a rho-independent terminator (section 7.5), the *attenuator*. If a cell contains adequate levels of the amino acid (no synthesis required) the leader peptide is synthesized and the terminator stops transcription. However, with inadequate levels of the amino acid (synthesis required), the ribosome forming the leader peptide 'stalls' when it reaches codon(s) specifying the given amino acid; a stalled ribosome causes the terminator region of the transcript to adopt an 'antiterminator' shape—and transcription of the operon proceeds.

Attenuator control occurs e.g. in the *his* (histidine) operon in E. *coli*.

7.8.2   Regulons

A regulon is a system in which two or more distinct genes and/or operons (each with its own promoter) are controlled by a common regulator molecule—all the genes/operons having similar initiator sequences. Examples in E. *coli* include catabolite repression and the SOS system.

7.8.2.1   Catabolite repression

Diauxic growth (section 3.4) shows that the presence of lactose does not *necessarily* induce the *lac* operon (section 7.8.1.1)—i.e. the effect of allolactose is overridden in the presence of glucose. Similarly, glucose will repress the *ara* operon (section 7.8.1.1) in the presence of arabinose. These are only two examples of *catabolite repression* (the 'glucose effect'): a common phenomenon (in bacteria) in which a cell uses some substrates in preference to others. The mechanism is complex and not fully understood; only an outline of one hypothesis is given below.

The (overall) regulatory molecule is the *catabolite activator protein* (CAP), also called CRP. To be functional, CAP must bind to (i.e. be activated by) cyclic AMP (cAMP) (Fig. 7.12) to form the cAMP-CAP complex.

cAMP-CAP allows the *lac, ara* and other operons to be expressed in the presence of their respective inducers. In the absence of cAMP-CAP, the RNA polymerase of e.g. the *lac* operon may bind to a site close to—but not at—the *lac* promoter; transcription from this alternative site could be weak, with only an occasional transcript being formed. When present, cAMP-CAP may displace the RNA polymerase, causing it to bind to the normal *lac* promoter.

The presence/absence of cAMP-CAP in the cell depends on the level of

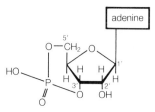

Fig. 7.12 Cyclic AMP (cAMP): a molecule involved e.g. in catabolite repression (section 7.8.2.1)—and, interestingly, in the pathogenesis of cholera (section 11.3.1). cAMP (adenosine 3′,5′-cyclic monophosphate) is synthesized from ATP by the enzyme adenylate cyclase, and is degraded to AMP by cAMP phosphodiesterase. A recent review deals with the various roles of cyclic AMP in prokaryotes [Botsford & Harman (1992) MR *56* 100–122].

cAMP. cAMP levels are low (cAMP–CAP absent) when glucose levels are high—and vice versa; why this is so is not understood. The level of cAMP seems to be determined mainly by the level/activity of its synthesizing enzyme (*adenylate cyclase*)—although another enzyme (cAMP phosphodiesterase) degrades cAMP to AMP; hence, to understand this system, we need to know how the presence of glucose affects the level/activity of the enzyme(s) governing the level of cAMP.

7.8.2.2   The SOS system in Escherichia coli

This system of genes is expressed when DNA is damaged and/or cannot replicate—due e.g. to the effects of ultraviolet radiation and/or certain chemicals: expression of the SOS system can e.g. stop cell division, increase DNA repair, affect energy metabolism and suppress restriction.

Control is exercised by the LexA protein (*lexA* gene product); normally, LexA binds close to the promoters of SOS genes and inhibits their transcription. Damage to DNA 'activates' the RecA protein (*recA* gene product) so that it causes breakdown of LexA—thus allowing expression of the SOS system; the mode of activation of RecA is unknown, but it may involve interaction between RecA and region(s) of single-stranded DNA.

The product of one SOS gene (*sulA*) represses *septum* formation (section 3.2.1), thus inhibiting cell division.

DNA repair (including excision repair: section 7.7.1) shows enhanced activity. A repair process involving genes *umuC* and *umuD* ('error-prone repair') operates only when the SOS system has been induced; the mechanism is unknown, but, as well as repairing DNA, it results in an increased number of *mutations* (Chapter 8).

Once DNA has been repaired, RecA is inactivated and LexA represses the SOS system.

7.8.2.3   Endospore formation/germination

Endospore formation (section 4.3.1) involves a range of proteins not synthesized in the vegetative (growing) cell, and this requires co-ordinated expression of various 'sporulation' genes; this topic is included under 'replicons' only tentatively as gene control in sporulation is still poorly understood. In *Bacillus subtilis*, some sporulation genes are apparently induced just before the stationary phase of growth; according to one scheme, the *spo0A* gene product—kept inactive by high levels of a particular metabolite—becomes active in lower levels of that metabolite and promotes the transcription of other genes. The early formation of several, new (sporulation-specific) sigma factors (section 7.5) permits transcription from new promoters; these factors (σ^{22} and σ^{29}) seem to be necessary for the completion of (at least) stage II of sporulation.

Mutant cells blocked in stage 0 cannot synthesize sporulation-specific products.

Several genes are known to be important in germination. For example, the *gerA* product is involved in the spore's response to L-alanine, and it may form part of a receptor for that germinant.

7.8.3   Recombinational regulation of gene expression

See site-specific recombination (section 8.2.2).

8 Molecular biology II: changing the message

DNA can change. For example, even while replicating—and despite proof-reading and repair systems (section 7.7)—about one in 10^8–10^{10} 'wrong' nucleotides are believed to be incorporated in the new (daughter) strand. Greater changes can be brought about by chemical and physical *mutagens* (section 8.1), by recombination (section 8.2), and by *transposable elements* (section 8.3). Additionally, the cell's *genome* (its 'genetic blueprint') can be supplemented by plasmids (section 7.1) and by other pieces of 'extra' DNA.

8.1 MUTATION

A *mutation* is a stable, heritable change in the sequence of nucleotides in DNA; note that a change in even a single *base-pair* (section 7.2.1) changes the sequence. Transcription of altered DNA produces altered RNA, and altered mRNA may specify a different polypeptide (Table 7.2) with different biological activity. Of course, mutations can affect control and recognition sequences as well as sequences encoding polypeptides.

Mutations occur spontaneously at low frequency without an obvious external cause; they are mainly errors in replication and repair, but chemical changes can also occur in DNA bases—for example, the deamination of cytosine to uracil (Fig. 7.3).

Mutation is encouraged by *mutagens*: physical agents such as ultraviolet radiation and X-rays, and chemicals such as alkylating agents, bisulphite, hydroxylamine, nitrous acid, and 'base analogues' (e.g. 5-bromouracil) which can substitute for normal DNA bases.

Mutagens work in various ways. Ultraviolet radiation can cause e.g. covalent cross-linking between adjacent thymines; correction of the resulting 'thymine dimers' by 'error-prone' repair (section 7.8.2.2) can generate a variety of mutations. Cross-linking can also occur with some alkylating agents. Bisulphites and nitrous acid can e.g. deaminate cytosine to uracil; although uracil is not a stable constituent of DNA (section 7.2), its different base-pairing specificity can cause the insertion of a different base (adenine instead of guanine) in the daughter strand at the next round of replication.

In a population of bacteria, mutations normally occur randomly, affecting different genes in different individuals. A cell in which a mutation has occurred is called a *mutant*. Mutations are often harmful—and may be lethal if the affected sequence of nucleotides encodes a vital product or function. Beneficial mutations include e.g. those which increase the cell's resistance to antibiotic(s). For example, a mutation may result in an altered ribosome such that streptomycin no longer binds to the ribosome and (therefore) does not inhibit protein synthesis; the (mutant) cell will thus exhibit resistance to this antibiotic.

A mutation giving increased fitness for growth under existing conditions may enable the (mutant) cell to outgrow other (non-mutant: *wild-type*) individuals in the population and become numerically dominant in that population; such 'natural selection' underlies the concept of *evolution*.

8.1.1   Types of mutation

Mutations occur in various ways, and they can have various effects on the genetic 'message'. Infrequently, a piece of DNA is lost, gained, inverted—or even *transposed* (section 8.3). A *point mutation* involves the loss, gain or substitution of a single nucleotide; even this, however, can have far-reaching consequences for the cell (Fig. 8.1).

8.1.2   The isolation of mutants

A *particular* mutation occurs spontaneously only at very low frequency in a population of bacteria; for example, within a population of E. *coli*, the loss of ability to ferment galactose occurs (on average) once every 10^{10} cell division cycles, i.e. a *mutation rate* of 10^{-10}. Even in populations treated with a mutagen, cells with a *particular* mutation are still greatly outnumbered by wild-type cells and by those with other types of mutation.

How can we isolate the one (or few) *specific* mutants from a large population of bacteria? If, for example, in a population of streptomycin-sensitive bacteria, a single cell has mutated to streptomycin resistance (section 8.1), that cell can grow on a solid medium containing streptomycin and can form a colony (section 3.3.1); all the other cells, which are inhibited by streptomycin, will not grow on such a medium. In general, this type of selective method can be used whenever the mutant can grow on a medium, or under conditions, which inhibit the growth of non-mutant cells.

However, suppose that (through mutation) a cell has lost the ability to synthesize a particular compound (e.g. amino acid) which is necessary for growth; such a metabolically dependent mutant (an *auxotroph*) can grow only if it is supplied with the appropriate compound. (The corresponding wild-type cell is called a *prototroph*.) How are auxotrophs isolated—given that any medium which allows an auxotroph to grow will also permit the growth of prototrophs? One method is to use a *minimal medium*, i.e. a medium containing the minimum nutrients needed by prototrophs; a given auxotroph can grow on minimal medium only if the medium has been supplemented with the auxotroph's specific growth requirement(s). If, for example, we wish to isolate a histidine-requiring auxotroph, minimal medium is supplemented with a *low* concentration of histidine—allowing limited growth of the auxotroph; a colony formed by an auxotrophic cell will soon exhaust the histidine in its vicinity (and therefore remain small) while the colonies of prototrophs reach a normal size. Small colonies are *presumed* to be those of auxotrophs and can be tested further.

In an alternative method for isolating auxotrophs, a low-density population containing both prototrophic and auxotrophic cells is inoculated

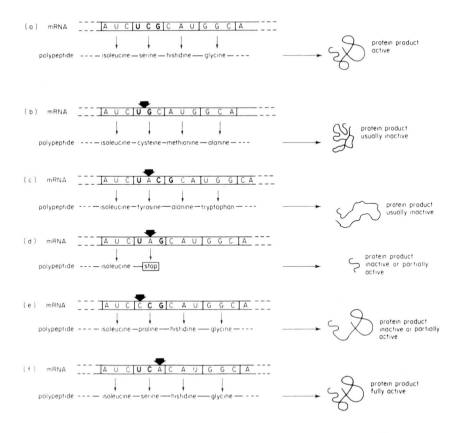

onto a *complete medium* (on which both prototrophs and auxotrophs can grow). Following incubation, normal-sized colonies are formed by all the cells, and in order to identify the colonies of auxotrophs it is necessary to inoculate each colony onto minimal medium—on which only the prototrophs will grow. This is conveniently achieved by *replica plating*. In this method a disc of sterile velvet is pressed gently onto the surface of the complete medium (the *master plate*) containing the colonies of both prototrophs and auxotrophs; cells from each colony stick to the velvet—which is then pressed lightly onto the surface of a sterile plate of minimal medium (the *replica plate*). After incubation, the positions of colonies on the replica plate are compared with those of colonies on the master plate; any colony which occurs on the master plate but not on the replica plate is presumed to be that of an auxotroph (Fig. 8.2).

Fig. 8.1 Point mutations: their effects on mRNA and on polypeptide synthesis. The effect of each point mutation is indicated by a heavy arrow (➡); at the right is shown a possible effect of each of the mutations.

(a) mRNA and polypeptide synthesized from the normal (non-mutant, wild-type) gene.

(b) The *deletion* of a guanine nucleotide from DNA has resulted in the loss of cytosine (C) from the codon UCG in mRNA. The effect of this is that not only UCG but all subsequent codons are altered: compare the amino acids encoded in (a) and (b). Notice that, if one nucleotide is missing, the next nucleotide is read in its place, i.e., groups of three consecutive nucleotides continue to be read as codons. Because the genetic message is out-of-phase 'downstream' of the deletion, such a mutation is called a *phase-shift* or *frame-shift mutation*; if it occurs near the end of a gene, so that most of the polypeptide is normal, the product may have some biological activity. If a phase-shift mutation occurs in an operon (section 7.8.1) the effect will vary greatly according to the particular site affected.

(c) The *addition* of a thymine nucleotide to DNA has resulted in the addition of an adenine nucleotide to codon UCG in the mRNA; as in (b), above, this is a phase-shift mutation.

(d) In DNA, thymine has replaced guanine, so that the mRNA now contains UAG (a 'stop' codon) instead of UCG; this is a so-called *nonsense mutation*. Polypeptide synthesis stops at UAG; the polypeptide may have some biological activity if much or most of it has been translated prior to UAG.

(e) In DNA, guanine has replaced adenine, so that the mRNA now contains CCG instead of UCG—the altered codon specifying proline rather than serine; this is a so-called *mis-sense mutation*. Note that the amino acids downstream of proline are not affected. The biological activity of the polypeptide will depend on the nature and position of the incorrect amino acid.

(f) In DNA, thymine has replaced cytosine, so that the mRNA now contains UCA instead of UCG; the altered codon still encodes serine (Table 7.2). This is a *silent mutation*; it does not, of course, affect the biological activity of the polypeptide.

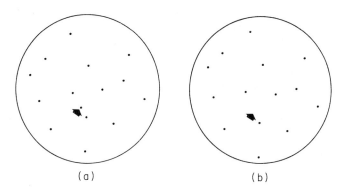

(a)   (b)

Fig. 8.2   Replica plating used for isolating an auxotrophic mutant. (a) Master plate: complete medium with colonies of both prototrophs and auxotrophs. (b) Replica plate: minimal medium with colonies of prototrophs only. On the master plate an arrow indicates a colony of a presumed auxotroph; there is no colony at the corresponding position (arrowed) on the replica plate (auxotrophs cannot grow on minimal medium).

8.1.3   The Ames test for carcinogens (*Salmonella*/microsome assay)

All carcinogens (cancer-promoting agents) are mutagens, and this test checks for potential carcinogenicity by checking for mutagenicity (the ability to cause mutations). The mutagenicity of a chemical is checked by determining its ability to reverse a previous mutation in the test organism, *Salmonella typhimurium*. (A mutation which reverses a previous mutation is called a *back mutation*.) The (mutant) test strains of *S. typhimurium* are auxotrophic for histidine, and the Ames test checks for back-mutation to prototrophy. Essentially, prototrophs are sought in an incubated mixture containing a population of the test strain, the chemical under test, and a preparation of enzymes from rat's liver; the enzymes are included because some mutagens/carcinogens need metabolic 'activation'.

8.2   RECOMBINATION

'Recombination' means the re-arrangement of one or more molecules of nucleic acid: molecules may e.g. join together, separate, or exchange strands, and a sequence within a molecule may be lost or inverted etc.

8.2.1   Homologous (general) recombination

Homologous recombination commonly involves strand exchange between two DNA duplexes. It can occur only if a long sequence of nucleotides in one of the duplexes is very similar to a sequence in the other, i.e. the two sequences must have extensive regions of *homology*; another requirement is the prior formation of one or more 'nicks' (breaks in the sugar–phosphate backbone) or 'gaps' (regions of single-stranded DNA). The RecA protein (product of the *recA* gene) has an essential role e.g. in strand exchange. In all cases a ligase joins 'new' and 'old' sections in affected strands.

Heteroduplex DNA (which has one strand from each parent duplex) may include mis-matched nucleotides; correction by mismatch repair (section 7.7) may use either strand as template.

Homologous recombination occurs e.g. in plasmid–chromosome interaction.

8.2.2 Site-specific recombination

In site-specific recombination (SSR), two specific sites in duplex DNA are brought together, a specific protein (a *recombinase*) binding at the juxtaposed sites; a staggered cut is made in each duplex, the cut ends are exchanged, and the strands are ligated (Fig. 8.3).

SSR can control gene expression; in such *recombinational regulation*, an SSR event controls the on/off switching of gene(s) or a switch from one gene to another (Fig. 8.3). SSR is also involved e.g. in the integration of bacteriophage λ DNA (section 9.2.1) with the *E. coli* chromosome (Fig. 8.3), and in the replicative form of transposition (Fig. 8.4).

8.3 TRANSPOSITION

Transposition is the transfer of a small, specialized piece of DNA—or a 'copy' of it—from one site to another in the same duplex, or to another site in a different duplex in the same cell. The 'small, specialized piece of DNA' is called a *transposable element* (TE; 'jumping gene'); TEs occur e.g. in bacterial chromosomes, in bacteriophage DNA (Chapter 9), and in plasmids.

The two main types of TE are the *insertion sequence* (IS) and the *transposon*. Each encodes protein(s)—including a *transposase*—which are needed for transposition; however, in addition to this, a transposon encodes other functions—e.g. enzyme(s) which inactivate particular antibiotic(s).

In all TEs, the nucleotide sequence includes so-called *inverted repeats*. An example of a pair of inverted repeats:

$$5'....CTGACTA.......TAGTCAG....3'$$
$$3'....GACTGAT.......ATCAGTC....5'$$

Notice that, when read in the 5'-to-3' direction, the sequence of nucleotides in the top strand is the same as that in the bottom strand. Inverted repeats

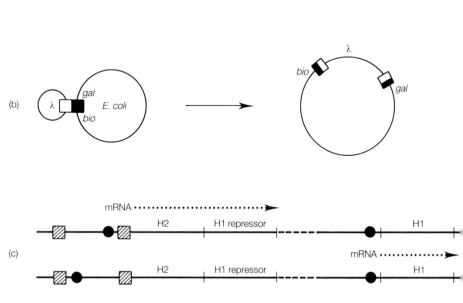

Fig. 8.3 Site-specific recombination: the principle (a) and some examples (b, c).
(a) A staggered break at one specific site in a DNA duplex; a staggered break involving the same nucleotide sequence is made at another site in the same duplex or in another DNA molecule. Each single-stranded region at a given breakage site can base-pair with a complementary region at the other breakage site. (The single-stranded regions are sometimes called 'sticky ends'.) The process is mediated by a protein (a *recombinase*) which recognizes a specific nucleotide sequence. The actual sequence in the sticky ends will depend on the particular recombinase involved.
(b) Integration of the (circular, double-stranded) DNA of bacteriophage λ (section 9.2.1) with the E. *coli* chromosome. In λ DNA, the specific recombinational site (i.e. the specific sequence of nucleotides in the duplex) is shown as a white square; in the E. *coli* chromosome it is shown as a black square flanked on either side by the *gal* gene (galactose utilization) and the *bio* gene (biotin synthesis). Left: the two circular DNA duplexes are shown with their recombinational sites juxtaposed. Initially, with the recombinase bound at the juxtaposed sites, a staggered break is made across each duplex; sticky ends in the chromosome then base-pair with those in the λ duplex, and the strands are ligated. Right: λ DNA incorporated in the E. *coli* chromosome.
(c) An example of recombinational regulation. In most strains of *Salmonella* the flagellar filament (section 2.2.14.1) can be made of either of two distinct types of protein—encoded by genes H1 and H2; normally, only one of these genes is expressed at any given time, so that the flagellar filament is made of either the H1 gene product *or* the H2 gene product. The promoter of the H2 operon (●) is flanked

by two specific sites (▨) that are recognized by a recombinase. Top: the H2 operon is transcribed; the H2 gene product forms the flagellar filament, and the H1 repressor stops transcription of the H1 gene. Bottom: site-specific recombination has occurred between the sites recognized by the recombinase, and the sequence containing the H2 promoter has been inverted; without a functional promoter, the H2 operon cannot be transcribed, but the loss of the H1 repressor permits transcription from H1—so that the filament is now made from the H1 gene product.

Variation in the composition of the *Salmonella* flagellar filament is only one example of *phase variation*: a more general phenomenon in which the composition of certain cell-surface components or structures undergoes spontaneous change. For example, in individual cells of *E. coli*, the so-called type 1 fimbriae are subject to on/off switching—resulting in spontaneous changes from a fimbriate to an afimbriate state, and vice versa; this, also, appears to be due to DNA re-arrangement.

---

seem to be necessary for transposition, probably being recognized by the appropriate transposase.

An *insertion sequence* consists of a pair of inverted repeats—each about 10-40 nucleotides long—bracketing the transposition genes.

A *class I transposon* (compound or composite transposon) consists of a pair of insertion sequences bracketing the gene sequence; example: transposon Tn*10*.

A *class II transposon* (simple transposon) consists of a pair of inverted repeats bracketing the gene sequence; example: transposon Tn*3*.

A TE may transfer by 'simple' transposition or by 'replicative' transposition (explained in Fig. 8.4); for some TEs the 'target' site is highly specific, but other TEs can insert almost at random.

Transposition is usually a rare event. It can result in a variety of re-arrangements—including insertions, deletions and inversions.

8.4   GENE TRANSFER

New DNA may enter a bacterium through *transformation, conjugation* or *transduction*; transduction (requiring bacteriophages) is covered in Chapter 9.

8.4.1   Transformation

In transformation, a bacterium (or protoplast) takes up from its environment a piece of DNA; this *transforming* (or *donor*) DNA may be chromosomal DNA (from a lysed cell) or a plasmid etc. Transformation of the recipient bacterium occurs if/when donor DNA integrates with its chromosome or (in some cases) develops as an independent molecule within the cell. Transformation occurs naturally e.g. in species of *Bacillus, Haemophilus* and *Streptococcus*—but in *E. coli* (and many other Gram-negative bacteria) it occurs only under artificial conditions (section 8.4.1.1). The following refers to natural transformation.

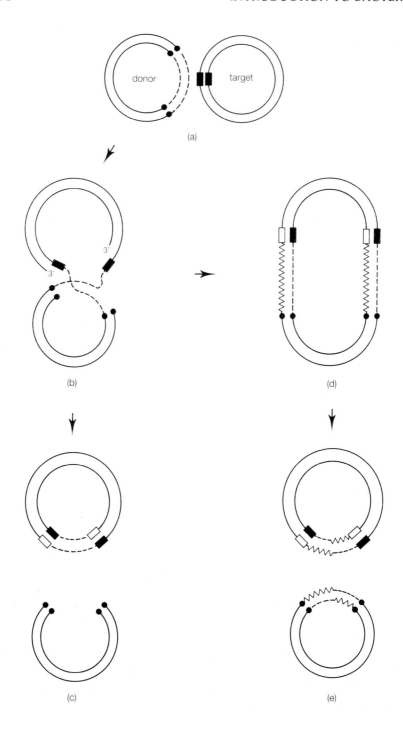

(a)

(b)   (d)

(c)   (e)

Cells which are able to take up DNA are said to be *competent*; competence occurs only in certain phase(s) of growth, and competent cells accept only double-stranded DNA above a certain minimum size. Initially, DNA binds at the cell surface. In *Bacillus subtilis* and *Streptococcus pneumoniae*, an enzyme/protein system degrades one strand and internalizes the other; the internalized strand quickly binds certain cytoplasmic proteins to form an *eclipse complex*—thus avoiding restriction (section 7.4). The single donor strand then replaces a similar (homologous) section in the recipient's DNA, e.g. by homologous recombination. In *Haemophilus*, both strands of donor DNA are apparently internalized, although only one strand is used.

Transformation was first observed by Griffith in the 1920s: a live, non-pathogenic strain of *S. pneumoniae* was found to become virulent when mixed with a dead, virulent strain; it was later found that DNA, released by the dead cells, had transformed the living ones—an early indication of the role of DNA as the carrier of genetic information.

Fig. 8.4 Transposition: a (diagrammatic) scheme for a transposon undergoing 'simple' and 'replicative' transposition.

(a) Two circular, double-stranded DNA molecules. The donor molecule includes a transposon (dashed lines)—either side of which is an old 'target' site (●); the donor's target site was duplicated when the transposon was originally inserted into the donor molecule (see later). The target molecule has a single target site (■) where the transposon will be inserted.

(b) A staggered break has been made at the target site. A nick has been made in each strand of the transposon (at opposite ends), and the free ends have been ligated to the target molecule, as shown. In 'simple' transposition (which occurs e.g. in transposon Tn*10*) the next (and final) stage is shown at (c).

(c) The result of 'simple' transposition. DNA synthesis has occurred from each free 3′ end in the target molecule, using the (single-stranded) target site as template, to form the complementary strand of each target site (▢). The remaining strand-ends of the transposon have been nicked and ligated as shown. Note that the target molecule's target site has been duplicated—compare with the donor molecule at (a). The rest of the donor molecule may be non-viable ('donor suicide').

(d) 'Replicative' transposition (which occurs e.g. in transposon Tn 3) involves stages (b), (d) and (e). At (d) new DNA synthesis (from each 3′ end in the target molecule) has continued beyond the target site, using each strand of the transposon as template; that is, the transposon has been replicated. The end of each new strand has been ligated to a free strand-end in the donor molecule. The structure shown at (d) is a *cointegrate*. The next stage involves a *resolvase*: an enzyme encoded by the transposon. This enzyme 'resolves' the cointegrate by promoting site-specific recombination (section 8.2.2) at a site in each transposon, forming the molecules shown at (e).

(e) Donor and target molecules each contain a copy of the transposon; notice that, in this model, each contains parts of the original transposon (dashed lines) as well as newly synthesized DNA (zig-zag lines). (Modified from *Molecular Biology of the Gene*, 4th edn (1987), by J.D. Watson *et al.* (Menlo Park, California: Benjamin/Cummings Publishing Company), p. 336, with permission.)

8.4.1.1   Laboratory-induced competence in transformation

Competence can be induced (e.g. in *E. coli*) by certain procedures which increase the permeability of the cell envelope. For example, calcium chloride solution (approx. 50 mM, 0.2 ml) containing 10^8–10^9 washed, mid-log-phase *E. coli* cells, is chilled on ice, and a DNA suspension (10 μl) is added to give a final DNA concentration of approx. 0.2 μg/ml; after further chilling at 0 °C (15–30 min) the suspension is heat-shocked (42 °C/1.5–2.0 min) and allowed to recover—e.g. returned to ice, then incubated in Luria–Bertani broth (1 ml) at 37 °C for 1 hour. (LB broth contains (per litre): 10 g tryptone, 5 g yeast extract and 10 g NaCl; pH 7.5, adjusted with NaOH.)

Small, circular plasmids tend to transform more readily than do larger ones. In *E. coli*, linear dsDNA transforms poorly (if at all) as it is degraded by the RecBC enzyme; however, it can transform some *recBC* mutants (which lack the enzyme).

8.4.2   Conjugation

Certain (*conjugative*) plasmids (section 7.1) confer on their host cells the ability to transfer DNA to other cells by *conjugation*; in this process, a *donor* ('male') cell transfers DNA to a *recipient* ('female') cell while the cells are in physical contact. A recipient which has received DNA from a donor is called a *transconjugant*. Conjugation differs mechanistically in Gram-positive and Gram-negative bacteria.

8.4.2.1   Conjugation in Gram-positive bacteria

In e.g. *Bacillus* and *Streptococcus* species, potential recipients secrete small amounts of a substance (a *pheromone*) which causes donor cells to synthesize an adhesive cell-surface component; DNA is transferred between adherent donors and recipients.

8.4.2.2   Conjugation in Gram-negative bacteria

In Gram-negative donors the plasmid encodes (among other proteins) the protein subunits of the *pilus* (section 2.2.14.3); pili seem to be essential for conjugation in Gram-negative bacteria. Different types of pilus promote conjugation under different physical conditions, and they seem to function in different ways.

One well-studied type of conjugation is that involving the F *plasmid* in *E. coli* host cells. Within each donor, the F plasmid usually exists as an independent circular molecule; donors, which bear pili, are designated F^+ cells, while recipients (which lack the F plasmid) are F^- cells. On mixing F^+ and F^- cells, the tips of the pili bind to the surfaces of F^- cells (see Plate 4:

bottom, right). What happens next is still disputed. Recently, American workers have claimed that DNA passes *through* the pilus to the F⁻ cell [Harrington & Rogerson (1990) JB *172* 7263-7264]. However, a different conclusion was reached when Swiss scientists subsequently observed conjugating cells by video-enhanced light microscopy, and examined donor-recipient contacts by electron microscopy [Dürrenberger, Villiger & Bächi (1991) JSB *107* 146-156]. Light microscopy showed donor and recipient cells rapidly drawn together (within a few minutes), and close wall-to-wall contact maintained for the next 80 minutes; the rapid development of wall-to-wall contact would involve pilus retraction, a feature widely accepted for some years. Specific 'conjugational junctions' at juxtaposed cell envelopes were seen by electron microscopy (Plate 4: top). From these observations, and from the kinetics of DNA transfer, the Swiss workers concluded that "... DNA is transferred at the state of close wall-to-wall contact rather than ... via extended pili.".

At some stage of contact, an (unknown) mating signal triggers DNA transfer. This starts with a nick at a specific site (*OriT*) in a specific strand of the F plasmid; the nick appears to be made by a plasmid-encoded protein, *helicase I* [Matson & Morton (1991) JBC *266* 16232-16237], which then unwinds the duplex from the free 5′ end. Details of the route and mode of DNA transfer are currently unknown. However, only the nicked strand enters the recipient—within which it acts as a template for DNA synthesis; hence, a complete, circular, copy of the plasmid forms in the recipient, which thus becomes F⁺. Within the donor, the lost strand is replaced by DNA synthesis, which may proceed according to the *rolling circle* model (Fig. 8.5).

Infrequently, an F plasmid integrates with the host cell's chromosome. When this happens, the fusion of plasmid with chromosome somewhat resembles the integration of phage λ (Fig. 8.3) in its overall effect—though the mechanism is different. The result is an *Hfr donor*. Hfr donors form pili, and they can conjugate with F⁻ cells. During Hfr × F⁻ crosses, DNA transfer starts (as in F⁺ donors) with a nick at *OriT* (see above). The transferred strand begins with plasmid DNA, but this is followed by chromosomal DNA and, finally—if the strand doesn't break—by the remainder of the plasmid strand. (This can be understood more easily if *OriT* is imagined to be in the middle of the integrated F plasmid.) If the entire plasmid and chromosomal strands are transferred the recipient becomes an Hfr donor, but usually strand breakage occurs at some point and the recipient (which does not receive *all* of the plasmid strand) remains F⁻. In both cases, however, the recipient generally receives *some* chromosomal (as well as plasmid) DNA, and if donor genes recombine with the recipient's chromosome they may alter the genetic message; the high proportion of recombinant cells resulting from Hfr × F⁻ crosses accounts for the designation 'Hfr'—i.e. 'high frequency of recombination'.

An F plasmid can also leave the chromosome, i.e. an Hfr donor can become an F⁺ donor. Sometimes, when leaving, the plasmid takes with it an adjacent piece of the chromosome; the result is an F' (F-prime) plasmid. In F' × F⁻ crosses, the F' donor transfers donor characteristics to the recipient (as does an F⁺ donor) but it also transfers chromosomal DNA (as does an Hfr donor).

The F plasmid is only one of many types of plasmid, and it is not even 'typical'; for example, the ability to form Hfr donors is not common. Also, in many types of plasmid the pilus-encoding and other 'donor' genes are normally repressed, i.e. in a population of potential donors only a few cells with transiently de-repressed plasmids can actually conjugate; by contrast, F⁺ cells are de-repressed, and in a population of them most or all can usually act as donors. As mentioned earlier (section 7.1), plasmids can encode—and may transfer—a range of functions.

'Universal' and 'surface-obligatory' conjugation. Plasmids which encode long, flexible pili—as does the F plasmid (section 2.2.14.3)—promote so-called 'universal' conjugation; this can occur equally well in a liquid medium (e.g. a broth culture) or on a moist, *non-submerged* solid surface (e.g. an agar plate). Experiments in liquid media indicate that 'universal' conjugation can be enhanced by raising the concentration of electrolyte in the medium [Singleton (1983) FEMS 20 151–153].

Plasmids which encode short, rigid, nail-like pili promote so-called 'surface-obligatory' conjugation—which occurs only on moist, *non-submerged* solid surfaces or in foams. This type of conjugation seems to require that conjugating cells be present within thin films of liquid—films which are thinner than the cells themselves; in nature, cells experience such conditions, for example, when present on soil particles that are drying by evaporation. Experiments have been carried out to see whether suitable conditions exist beneath the thin end of a liquid meniscus on chemically clean glass (Plate 4: bottom left); the results suggest that 'surface-obligatory'

Plate 4.   *Top.* Conjugating cells of *Escherichia coli* in wall-to-wall contact (scale: 6 cm = 1 μm). The electronmicrograph shows a recently discovered 'conjugational junction' (see text): the electron-dense (dark) line (between the arrowheads) which marks the region of contact between donor and recipient. As well as conjugating, the cell at the top is about to divide; compare the conjugational junction with the site of cell division (far right). The masses of small, darkly stained 'dots' are ribosomes. *Bottom right.* Conjugating cells of *E. coli*; this preparation has been stained to show an F pilus (arrowheads) passing between the cells. (Fragments of flagella are also visible.) *Bottom left.* An apparatus used in the author's experiment to investigate 'surface-obligatory' conjugation (see text): a bundle of 80 chemically clean 73-mm glass capillary tubes (internal diameter approx. 0.8 mm) within a universal bottle. When shaken up-and-down, the mating medium forms a large number of small 'threads' of liquid in the capillary tubes, each thread having a meniscus at both ends.
Photographs of *E. coli* courtesy of Dr Markus B. Dürrenberger, University of Zürich, Switzerland.

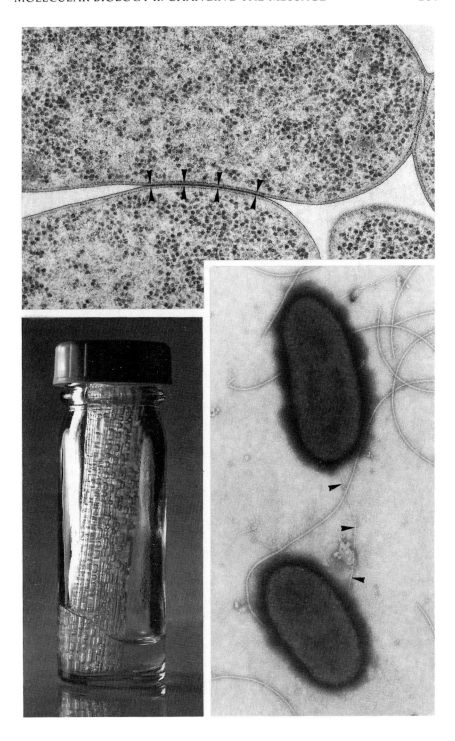

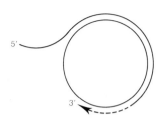

Fig. 8.5   The rolling circle model of DNA synthesis in a circular, double-stranded molecule. First, a nick is made in one strand. Then, using the un-nicked strand as template, a DNA polymerase extends the 3′ end of the nicked strand by adding nucleotides in the 5′-to-3′ direction—the 5′ end being progressively displaced. The displaced strand may itself be used as a template for the formation of Okazaki fragments (Fig. 7.8).

conjugation requires, or is assisted by, surface tension [Singleton (1983) FEMS 19 179–182, (1983) JGM 129 3697–3699].

8.5   GENETIC ENGINEERING

The genome of a bacterium can be modified in the laboratory; certain bacteria have been altered in this way so that they can make useful products for industry, agriculture and medicine.

Some bacteria—and/or their enzymes and plasmids—are also widely used for manipulating DNA; to illustrate a few *principles*, some common techniques are outlined below.

8.5.1   Cloning (molecular cloning)

Cloning is a method for obtaining many copies of a gene (or other piece of DNA)—e.g. for analysis, for making probes (see later), or for large-scale production of the gene product. Initially, the gene is isolated and is then inserted into a so-called *vector* molecule, such as a plasmid (section 7.1), as shown in Fig. 8.6. The (hybrid) plasmid (Fig. 8.6) can then be inserted into a bacterium—e.g. by transformation (section 8.4.1)—so that the gene is copied each time the plasmid replicates; such bacteria can be grown to high numbers, thus producing many plasmids and (hence) many copies of the particular ('cloned') gene. To harvest the gene, the bacteria are lysed; certain reagents, added during the process, permit separation of nucleic acids (chromosomal and plasmid DNA, and RNA) from the rest of the cell debris. The plasmids can then be isolated by specialized centrifugation. By using the original restriction endonuclease (Fig. 8.6), the cloned gene can be cut from each plasmid and separated (from plasmid DNA) e.g. by electrophoresis on agarose gel.

Phages (Chapter 9) can also be used as vectors; larger pieces of DNA can be cloned in these vectors.

8.5.2 Probes

Suppose that we wish to find out whether a given type of plasmid or gene etc. contains a specific short sequence of nucleotides. One way is to use a *probe*: a single-stranded piece of DNA of the required sequence which has been 'labelled' in some way. If a suspension of plasmids and probes is heated, the strands of the (double-stranded) plasmid DNA separate; on cooling, some of the probes will base-pair with their complementary sequence (if present) before the plasmid DNA has re-formed. If free (non-duplexed) probes are then removed, any remaining probes—detected by their 'label'—will indicate the presence of the given sequence in the plasmid. Probe DNA which has been labelled with ^{32}P, for example, is detected by its radioactivity.

If the given sequence is found, its location in the plasmid may be determined by cutting up the plasmid (with restriction endonucleases) and examining each piece in turn.

Probes can be made e.g. by cloning the particular sequence (as in section 8.5.1). Before excision from the hybrid plasmid (Fig. 8.6), the probe DNA can be labelled by so-called *nick translation*, as follows. An endonuclease makes a few single-strand nicks in each hybrid plasmid; then, another enzyme (e.g. the *E. coli* DNA polymerase I) removes nucleotides from nicked strands and replaces them with labelled nucleotides which are present (in excess) in the reaction mixture. Probe DNA (labelled as part of the hybrid plasmid) can then be isolated (by restriction endonucleases) and the strands separated by heating.

8.5.3 The polymerase chain reaction (PCR)

PCR is a rapid method for making many copies of a given piece of (double-stranded) DNA; it depends on the ability of a DNA polymerase to extend a primer (section 7.3, Fig. 7.8). The reaction mixture contains (i) the sample DNA; (ii) the primers: 15–20-nucleotide pieces of single-stranded DNA which are complementary to the 3' end of each strand of the sample duplex; (iii) the *Taq* DNA polymerase; (iv) deoxyribonucleotide triphosphates. On

Table 8.1 Restriction endonucleases: some common examples

Restriction endonuclease	Source	Recognition sequence and cutting site[1]
EcoRI	Escherichia coli	G/AATTC
BamHI	Bacillus amyloliquefaciens	G/GATCC
HindIII	Haemophilus influenzae	A/AGCTT
PstI	Providencia stuartii	CTGCA/G
BclI	Bacillus caldolyticus	T/GATCA

[1]The solidus (/) indicates the cutting site. The sequence is written in the 5'-to-3' direction.

heating the mixture to 95 °C, the strands of the (sample) duplex separate, and transient cooling to 37 °C allows primers to bind at the 3′ end of each strand. On heating to 72 °C, *Taq* polymerase extends each primer by adding nucleotides in the 5′-to-3′ direction; two DNA duplexes are thus formed from the original duplex. The process is repeated a number of times, the yield of sample DNA increasing each time.

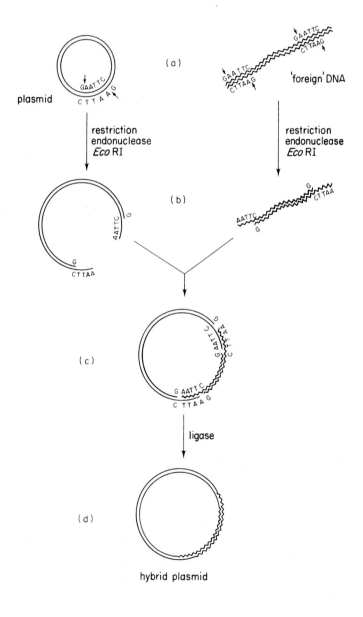

The key component in PCR, the heat-stable *Taq* polymerase, permits the process to be carried out at a temperature at which the template strands do not readily re-combine—strand separation being necessary for DNA replication (section 7.3). The *Taq* enzyme is from *Thermus aquaticus*, a bacterium which lives in hot springs and which has an optimum growth temperature (section 3.1.4) of about 66–75°C.

Fig. 8.6 Construction of a hybrid plasmid. (a) A plasmid (isolated from a bacterium), and 'foreign' DNA e.g. from another bacterium or from a eukaryotic cell. Both molecules include the nucleotide sequence GAATTC which is recognized by the restriction endonuclease *Eco*RI (section 7.4); *Eco*RI cuts between nucleotides G and A, as shown by the arrows. (b) As a result of *Eco*RI activity, both molecules have 'sticky ends' (terminal, single-stranded complementary sequences of nucleotides) which allow integration by base-pairing—as shown at (c). (d) An enzyme, a DNA ligase, has catalysed a phosphodiester bond (Fig. 7.4) between the sugar residues of each G and A nucleotide; the hybrid plasmid thus formed can, if required, be inserted into a bacterium for cloning.

Different segments of DNA can be inserted into different regions of a plasmid by the use of different restriction endonucleases—provided, of course, that each molecule has appropriate restriction site(s) (see e.g. Table 8.1).

9   Bacteriophages

Most or all bacteria can be infected by specialized viruses (*bacteriophages*, usually abbreviated to *phages*). A *virus* is an organism which does not have a cell-type structure and which cannot, by itself, metabolize or reproduce. However, when inside a suitable living cell the genome of a virus may 'take over' the synthesizing machinery and direct it to make copies of the virus; the newly-formed viruses are released and can then infect other cells.

Many phages consist simply of nucleic acid enclosed within a protein *capsid* ('coat'); depending on phage, the genome may be dsDNA, ssDNA, dsRNA or ssRNA. Some phages are polyhedral, others filamentous or pleomorphic, and some have a 'tail' with which they attach to the host cell (Table 9.1, Fig. 9.1, Plate 5). In many cases a given phage can infect the cells of only one genus, species or strain.

The effect of phage infection depends on the particular phage and host cell—and, to some extent, on conditions. *Virulent* phages multiply within and *lyse* their host cells, i.e. the host cells die and break open, releasing phage progeny. *Temperate* phages can establish a stable, non-lytic relationship (*lysogeny*) with their host cells. Still other phages can multiply within their hosts without destroying them, phages being released from the living cells.

9.1   VIRULENT PHAGES: THE LYTIC CYCLE

The lytic cycle begins when a virulent phage adsorbs to a susceptible host cell; it ends with cell lysis and the release of phage progeny.

9.1.1   The lytic cycle of phage T4 in *Escherichia coli*

A phage attaches by its tail fibres to specific sites on the outer membrane; subsequently the base-plate binds to the cell surface. The phage sheath then contracts, the inner core (Fig. 9.1) penetrates the outer membrane, and phage DNA passes via the central duct into the periplasmic region. Uptake of DNA across the cytoplasmic membrane seems to require the presence of pmf (section 5.1.1.2).

Within 5 minutes of infection, the host cell stops making its *own* DNA, RNA and protein. Some of the T4 genes (the 'early' genes) are transcribed and translated (section 7.6), and T4-encoded nucleases start to degrade the host's chromosome—releasing nucleotides which are used later for the synthesis of T4 DNA.

Table 9.1 Bacteriophages: some examples

Name	Genome[1]	Morphology[2]; size[3]	Main host(s)
λ	dsDNA (linear)	Isometric head, long non-contractile tail; ca. 200 nm	*Escherichia coli*
Mu	dsDNA (linear)	Isometric head, long contractile tail; ca. 150 nm	enterobacteria
MV-L3	dsDNA (linear)	Isometric head, short tail; ca. 80 nm	*Acholeplasma laidlawii*
T4	dsDNA (linear)	Elongated head, long contractile tail; ca. 200 nm	*Escherichia coli*
PM2	dsDNA (ccc)	Icosahedral, with internal lipid membrane; ca. 60 nm	*Alteromonas espejiana*
f1	ssDNA (ccc)	Filamentous; > 750 nm × 6 nm	enterobacteria (only conjugative *donor* cells can be infected)
φX174	ssDNA (ccc)	Icosahedral; ca. 30 nm	*Escherichia coli*
M12	ssRNA (linear)	Icosahedral; ca. 25 nm	enterobacteria (only conjugative *donor* cells)

[1]The form in which the nucleic acid exists within the virus.
[2]Isometric means approximately spherical, commonly icosahedral; 'icosahedral' means resembling an *icosahedron*: a solid figure bounded by 20 plane faces, all the faces being equilateral triangles of the same size.
[3]Excludes tail fibres (where present).

Other ('medium', 'late') T4 genes are transcribed sequentially. Throughout, transcription seems to involve the host's RNA polymerase, but this enzyme undergoes a series of phage-induced changes which enable it to recognize promoters of the 'middle' and 'late' phage genes. One such change is *ADP-ribosylation*: the transfer, from NAD (Fig. 5.1) to the RNA polymerase, of an ADP–ribosyl group; additionally, T4 encodes a sigma factor (section 7.5) necessary for transcription of the T4 'late' genes.

About 5 minutes after infection, T4 DNA begins to replicate from more than one 'origin'; 'leading' strand synthesis (Fig. 7.8) may be primed by the cell's RNA polymerase, but Okazaki fragments are primed by T4-encoded proteins—which also carry out other functions. New T4 DNA undergoes modification (section 7.4).

'Late' genes encode various phage components. Assembly of the phage head involves at least 20 genes; *scaffolding proteins*, which form a 'prohead' on which the phage head is built, are later removed, and the head is subsequently filled with DNA. The 'tail' is polymerized (core first) on the base-plate, and the completed tail joins spontaneously to the DNA-containing head.

Release of phage progeny seems to involve osmotic lysis following

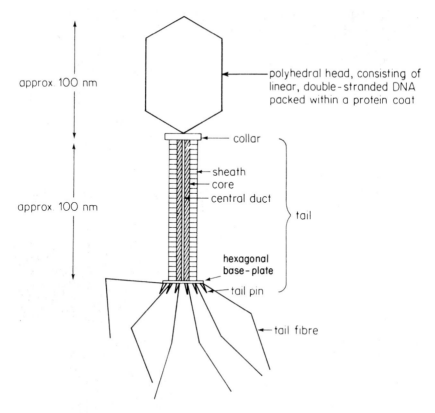

Fig. 9.1   Bacteriophage T4, a complex, tailed phage which infects *Escherichia coli* (not drawn to scale). The purely structural parts are made of protein; the genome (dsDNA) is enclosed within the 'head'. The tail fibres are shown extended, but they are normally folded back along the tail sheath and head; extended fibres are needed for infection. Extension of tail fibres is inhibited e.g. by low pH, low temperature and low ionic strength. 1 nm (nanometre) = $10^{-3}\,\mu$m = 10^{-6} mm.

Plate 5.   Some bacteriophages (see text and Table 9.1). *1.* Bacteriophage λ. *2* and *3.* Bacteriophages T6 and T4, respectively (both to the same scale). *4.* Bacteriophage fd, a filamentous phage containing ss cccDNA (bar = 50 nm). *5.* Bacteriophage φ29. The elongated head (approx. 35–40 nm), which contains linear dsDNA, bears protein fibres and is connected to a short, non-contractile tail. *6.* Bacteriophage Qβ, a small icosahedral phage (approx. 25 nm in diameter) which contains ssRNA. *7* and *8.* Bacteriophages T3 and T7, respectively (both to the same scale: bar = 50 nm); both phages contain linear dsDNA, and each has a short, non-contractile tail. *9.* Bacteriophage T1 (bar = 50 nm); the head contains dsDNA, and the long tail is non-contractile.

Courtesy of Dr Michel Wurtz, University of Basle, Switzerland, and reprinted, with permission, from selected micrographs in 'Bacteriophage structure' by Dr M. Wurtz, *Electron Microscopy Reviews* vol. 5(2) (1992), Pergamon Press plc.

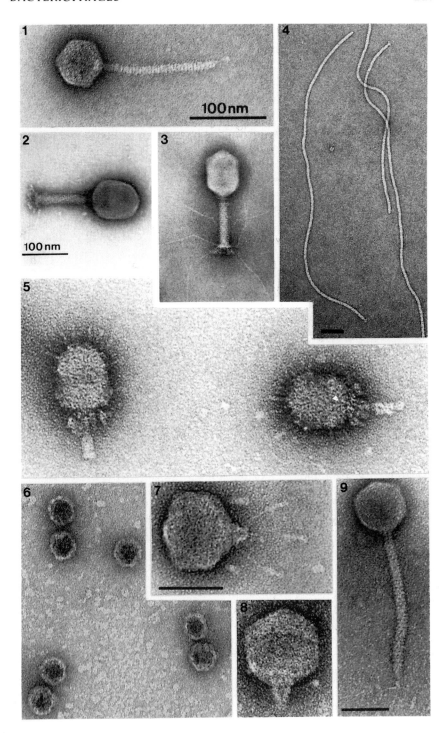

weakening of the cell envelope by T4-encoded *lysozyme* (see caption to Fig. 2.7).

9.1.2   Lytic cycles of other phages

Lytic cycles can differ markedly from that of T4—as shown in the following examples. (i) Not all phages bind initially to the cell wall; some bind e.g. to specific sites on particular types of pilus—M12, for example, binds to the side (not the tip) of an F or F-type pilus. (ii) In some cases the host's chromosome is allowed to survive for some time so that certain host genes can be used for phage replication. (iii) In small ssRNA phages such as M12, MS2 and Qβ, the phage genome itself acts as mRNA, i.e. only translation of the phage genome is required. For replication, RNA phages must encode at least part of the replicase system since no bacterium can synthesize RNA on an RNA template. (iv) Some phages encode a specific 'lysis protein' which promotes cell lysis—and (hence) release of phage progeny; the lysis protein of phage MS2 seems to work by causing the formation of specific adhesion sites in the cell envelope [Walderich & Höltje (1989) JB *171* 3331–3336].

9.1.3   The effect of virulent phages on bacterial cultures

If virulent phages are added to a broth culture of susceptible bacteria, most or all of the cells will subsequently lyse; this can cause a dense, cloudy culture to become clear.

If a *small* number of virulent phages is added to a *confluent* layer of susceptible bacteria (section 3.3.1), each individual phage will infect and lyse a single cell, releasing many phage progeny; progeny phages can then infect and lyse neighbouring cells, and so on. In this way, a visible, usually circular, clearing—called a *plaque*—develops in the opaque layer of confluent growth at each site where one of the original phages infected a cell.

9.2   TEMPERATE PHAGES: LYSOGENY

A *temperate* phage can enter into a stable, non-lytic relationship with a bacterium; the relationship is called *lysogeny*, and the host bacterium is said to be *lysogenic*. A lysogenic bacterium is immune from attack by other phages of similar type (*superinfection immunity*). In most cases of lysogeny the phage genome (i.e. its nucleic acid)—called the *prophage*—integrates with the host's chromosome; in a few cases (e.g. phage P1) the prophage does not integrate with the chromosome but exists within the bacterium as a circular molecule ('plasmid'). Either way, replication of the prophage and host cell is co-ordinated so that, at cell division, each daughter cell receives a phage.

Maintenance of the lysogenic state involves phage-directed synthesis of a repressor protein. Loss of active repressor protein results in *induction* of the lytic cycle—i.e. a prophage retains the potential for virulence. In a population of lysogenic bacteria induction may occur spontaneously in a small number of cells.

Lysogeny seems to be common in nature.

9.2.1   Lysogeny of phage λ in *Escherichia coli*

DNA enters the bacterium, via the phage's tail, as *linear* dsDNA (Table 9.1); it has 'sticky ends' (explained in Fig. 8.6) which immediately base-pair, and are ligated, so that the phage DNA circularizes.

While the 'early' phage genes are being transcribed either lysogeny or lysis may follow. The lysis/lysogeny decision depends on the 'success' of one or other of two early gene products: the cI protein (which represses transcription from certain operators, and which promotes lysogeny) and the cro protein, which inhibits transcription of the *cI* gene. The cell's internal state can affect this decision. For example, lysogeny is favoured by starvation; one explanation of this is that starvation increases levels of cAMP (Fig. 7.12), a factor believed to favour synthesis of the cI protein.

In lysogeny, the λ prophage integrates with the host's chromosome by site-specific recombination (Fig. 8.3b); a phage-encoded protein (product of the *int* gene) acts as the specific recombinase.

Induction (the switch from lysogeny to virulence) occurs spontaneously in a few cells in a lysogenic population. Induction in most or all of the cells can be brought about e.g. by DNA-damaging agents such as ultraviolet radiation or the antibiotic mitomycin C; under these conditions, when the SOS system is operative (section 7.8.2.2), the activated RecA protein cleaves the phage repressor protein (cI protein), allowing expression of the lytic cycle. This requires *excision* of the prophage—the reverse of integration; excision is mediated by the phage-encoded xis protein. The prophage is then replicated, phage components are synthesized, and the assembled phages are released on cell lysis.

9.3   ANDROPHAGES

Androphages infect only certain bacteria which contain a conjugative plasmid. For example, f1 and fd are filamentous, ssDNA phages (Table 9.1) which adsorb specifically to the *tips* of certain types of pili; penetration of the host cell may involve pilus retraction. The progeny phages of f1 and fd are released through the cell envelope; host cells remain viable, but they grow more slowly than do uninfected cells.

M12, MS2, f2 and Qβ are small, icosahedral ssRNA phages (Table 9.1) which adsorb to the *sides* of certain types of pili; MS2 encodes a 'lysis protein' (section 9.1.2).

9.4   PHAGE CONVERSION

Bacteria infected with phage may have certain characteristics not shown by uninfected cells; such *phage conversion* (= bacteriophage conversion) may be due e.g. to expression of phage genes by the cells, or to inactivation of chromosomal genes through integration of the prophage. For example, strains of *Corynebacterium diphtheriae* which cause diphtheria are lysogenized by a certain type of phage which encodes and expresses a potent toxin; strains which lack the phage cannot form the toxin and do not cause diphtheria. In *Staphylococcus aureus*, integration of the prophage of phage L54a causes loss of lipase activity due to inactivation of the relevant chromosomal gene. (See also *Salmonella* in the Appendix.)

9.5   TRANSDUCTION

The transfer of chromosomal (or plasmid) DNA from one cell to another via a phage is called *transduction.*

9.5.1   Generalized transduction

In this process, any of a variety of genes may be transferred from one cell to another. In a population of phage-infected bacteria it occasionally happens that, during phage assembly, chromosomal or plasmid DNA is incorporated in place of phage DNA; such abnormal phages can, once released, attach to other cells and donate DNA but (as it is not phage DNA) neither lysogeny nor lysis will result.

In the recipient cell (*transductant*) the transduced DNA may (i) be degraded by restriction endonucleases (section 7.4); (ii) undergo recombination with the chromosome (or plasmid) so that some donor genes can be stably inherited (*complete transduction*); (iii) persist as a stable but non-replicating molecule (*abortive transduction*). (If an abortive transductant gives rise to a colony, only one cell in the colony will contain donor DNA.)

Donor genes with functional promoters (whether in a complete or abortive transductant) may be expressed in the transductant.

The transfer of any given gene by generalized transduction is a rare event. If two or more genes are transduced together (*co-transduction*), this is taken as evidence that they occur on the same fragment of DNA; such information has been useful for the detailed mapping of donor chromosomes and plasmids—distances between genes being estimated from co-transduction frequencies.

9.5.2 Specialized (restricted) transduction

Specialized transduction can be brought about only by a temperate phage which integrates with the host's chromosome (see e.g. section 9.2.1). On excision, a prophage will occasionally take with it some of the adjacent chromosome—an event similar (in principle) to the formation of an F' plasmid (section 8.4.2.2). In the $E.\,coli$/phage λ system, the integrated prophage is flanked on either side by the host's gal and bio genes (Fig. 8.3b); hence, in one of the rare 'aberrant' excisions, the prophage may take with it either the gal or bio gene—often leaving behind certain phage genes from the opposite end of the prophage. Such a phage (a specialized transducing particle, or STP) may lack the ability to replicate (i.e. it may be *defective*); nevertheless, an STP can inject its DNA into a recipient cell and (hence) transfer specific donor genes (gal or bio). In other bacterium/phage systems, the genes flanking the integrated prophage are those which can be specifically transduced.

10   Bacteria in the living world

Bacteria are often thought of as pests to be destroyed, or as convenient 'bags of enzymes'—useful for experimental purposes. However, bacteria have a life of their own outside the laboratory, and many of their activities are important not only to man but also to the whole balance of nature. This aspect of bacteriology has many facets, and only a brief outline of some of them can be given.

10.1   MICROBIAL COMMUNITIES

Most bacteria are *free-living*, i.e. they do not necessarily form specific associations with other organisms; nevertheless, they are part of the web of life, and in nature they can rarely grow without affecting—or being affected by—other organisms. Bacteria normally occur as members of mixed communities which may include fungi, algae, protozoa and other organisms. Such communities can be found in a wide variety of natural habitats—e.g. in water, in soil, on the surfaces of plants, and on and within the bodies of man and other animals. Those microorganisms which are normally present in a particular habitat are referred to, collectively, as the *microflora* of that habitat.

Microorganisms which colonize a given habitat may affect each other in various ways; for example, they may have to compete for scarce nutrients, for oxygen, or for space etc., and those organisms which cannot compete effectively are likely to be eliminated from the habitat. In some cases an organism can actively discourage at least some of its competitors by producing substances which are toxic to them—a phenomenon termed *antagonism*; a microorganism which produces antibiotics (section 15.4), for example, may have a competitive advantage. There may also be relationships in which one or both organisms benefit and neither organism is harmed; for example, an acid-producing organism can help to create favourable conditions for another organism whose growth depends on a low pH.

If a habitat remains undisturbed there will eventually develop a stable community of organisms in which the various beneficial and antagonistic interactions have reached a delicate state of balance. An alien micro-organism will often have difficulty in establishing itself in such a community—unless a disturbance in the environment upsets the balance in the community. For example, in the intestine of an animal, the natural microflora can often discourage the establishment of a pathogen because (i) they occupy space (thereby hindering access), and (ii) they are well adapted to the intestine, and, for that reason, can usually outgrow a pathogen; however, any disturbance to the microflora—due e.g. to antibiotic therapy—may enable a pathogen to become established and cause disease.

10.1.1   Transient communities; cyanobacterial blooms

In contrast to the stable, mixed communities of microorganisms in many habitats, there are occasions when one, or a few, species transiently predominate.

In cholera (Chapter 11), the patient's intestine becomes a living incubator for the causal organism, *Vibrio cholerae*, and the so-called 'rice-water stools' may contain up to 10^9 cells/ml of *V. cholerae*.

In lakes, reservoirs and other bodies of water, certain conditions can encourage prolific growth of particular organisms. The result is a so-called *bloom*: a visible (often conspicuous) layer of organisms at or near the surface; the organisms include (or may consist mainly of) certain cyanobacteria—particularly those which form gas vacuoles (section 2.2.5)—and/or certain eukaryotic microorganisms. Blooms can be encouraged e.g. by an excess of nutrients (such as nitrogen leached from agricultural fertilizer) and/or by thermal stratification in the water. The death/ decomposition of the bloom-forming organisms (due e.g. to cessation of favourable factors) can cause a severe depletion of oxygen in the water, often resulting in the asphyxiation of fish and other aquatic animals.

In some cases bloom formation can be discouraged by pumping (circulating) the water to avoid stratification and/or by using anti-cyanobacterial chemicals such as *dichlone* (dichloronaphthoquinone).

In reservoirs, a substance (*geosmin*), produced by *Anabaena* and some other bloom-formers, imparts an 'earthy' or 'musty' taste to the water (and to fish living in the water).

Some bloom-forming cyanobacteria (e.g. species of *Anabaena, Microcystis, Nodularia*) produce toxins which can be lethal to fish and/or other animals.

10.2   SAPROTROPHS, PREDATORS, PARASITES, SYMBIONTS

10.2.1   Saprotrophs

Organisms which obtain nutrients from 'dead' organic matter are called *saprotrophs*. Some saprotrophs use only soluble compounds, but others can degrade cellulose (section 6.2) and other polymers, outside the cell, and assimilate soluble products. Complex substrates may be degraded in a stepwise manner, each of several species of saprotroph carrying out one (or a few) steps in the process; co-operation of this sort is important in the breakdown, and hence re-cycling, of organic matter in nature. Indeed, there are very few biological compounds which cannot be readily broken down by a community of saprotrophs working as a team. Without the (heterotrophic) saprotrophs the carbon cycle (Fig. 10.1)—and, hence, the other cycles of matter—would stop.

10.2.2   Predators

Members of the order Myxobacterales are Gram-negative rods which live in soil, on dung, and on decaying vegetation. Most species obtain nutrients by preying on other microorganisms: they release enzymes which lyse other bacteria, and fungi, and live on the soluble products. As predators, however, the myxobacteria are not typical: a predator usually ingests its prey *before* digesting it. In certain habitats bacteria themselves are prey for a wide variety of protozoan predators, an individual predator often consuming many thousands of bacteria in its lifetime; such protozoa can therefore greatly affect bacterial populations in these habitats.

Other predatory bacteria include *Bdellovibrio* (see Appendix).

10.2.3   Parasites

A *parasite* is an organism which lives on or within another living organism (the *host*) and which benefits (in some way) at the expense of the host; in almost all cases a parasite obtains nutrients from its host. The host may suffer varying degrees of damage—ranging from slight inconvenience to death.

Parasitism may be adopted as an alternative way of life by certain free-living bacteria, but in some bacteria it is obligatory. Bacteria such as *Mycobacterium leprae* (which causes leprosy) can grow only within particular types of (eukaryotic) host cell; obligate parasites such as this depend heavily on their host's metabolism, and often they cannot be grown in the laboratory except in specialized preparations of living cells.

A parasite which affects its host severely enough to cause disease, or death, is called a *pathogen*. Not all parasites are pathogens, and not all pathogens are parasites; an example of a non-parasitic pathogen is *Clostridium botulinum* (section 11.3.1).

10.2.4   Symbionts

Originally, *symbiosis* meant any stable, physical association between different organisms (the symbionts)—regardless of the nature of their relationship. Later, the meaning of the term was restricted to cover only those instances in which the relationship was one of mutual benefit. However (despite the potential for confusion) there is now a general tendency to move back to the original meaning. Hence, in the common understanding of symbiosis, the possible relationships between symbionts include mutual benefit (*mutualism*) and parasitism. Using the same terminology, a *commensal* is a symbiont which gains benefit from another symbiont such that the latter derives neither benefit nor harm from the association.

10.2.4.1 Mutualistic symbioses

Mutually beneficial relationships between bacteria and other organisms are quite common. For example, ruminants (e.g. sheep, cows etc.) cannot produce enzymes to digest the cellulose in their diet of plant material; however, in these animals the gut includes a specialized compartment (the *rumen*) containing vast numbers of microorganisms (including bacteria such as *Ruminococcus*) which convert cellulose to simple products that the animal can absorb. In return, the microbes benefit from a warm, stable environment and the abundance of nutrients which the animal swallows.

In certain insects, specialized cells (*mycetocytes*) in the gut lining contain (intracellular) bacteria which, in at least some cases, supply essential nutrients to the host. A *mycetome*, a distinct organelle composed of a group of mycetocytes, may be associated with the gut.

In leguminous plants (peas, beans, clover etc.) the roots have small swellings (*nodules*) containing bacteria of the genus *Rhizobium*; in this arrangement, the plant provides nutrients and protection, while *Rhizobium* supplies the plant with 'fixed' nitrogen from the atmosphere (section 10.3.2). Root nodules enable these plants to thrive in nitrogen-poor soils.

Nitrogen-fixing bacteria also form associations with non-leguminous plants. For example, nodules in the roots of the alder tree (*Alnus*) contain bacteria of the genus *Frankia*, and the small floating fern *Azolla* contains *Anabaena azollae* within specialized cavities.

10.3 BACTERIA AND THE CYCLES OF MATTER

The elements which make up living organisms occur on Earth in finite amounts; accordingly, for life to continue, the components of dead organisms must be re-used or 're-cycled'. Bacteria, together with other microorganisms, play a vital role in this process.

10.3.1 The carbon cycle

In all living organisms the major structural element is carbon (Chapter 6), so

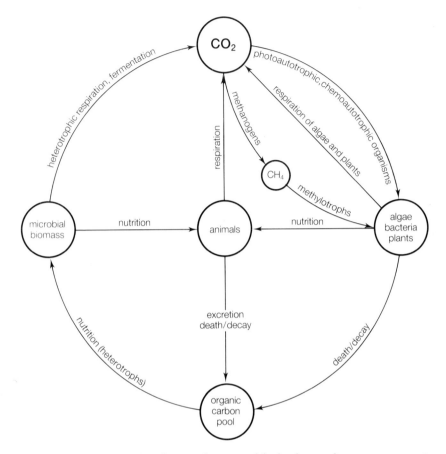

Fig. 10.1 The biological carbon cycle: a simplified scheme showing some major interconversions of carbon in nature. Bacteria have significant roles both as autotrophs and heterotrophs (Chapter 6), and unique roles as methanogens (section 5.1.2); the methylotrophs (section 6.4) include methane-utilizing bacteria. Microorganisms (including bacteria) are responsible for the essential conversion of 'dead' organic carbon to biomass and CO_2; without this process the cycle would stop. Microbial biomass, as such, is used e.g. by filter-feeders (oysters etc.) and, via food chains, by fish and other animals. The role of *elemental* carbon seems to be minimal (compare with nitrogen and sulphur in their respective cycles).

that the re-cycling of carbon is of fundamental importance. In the biological carbon cycle (Fig. 10.1) the chief contribution of bacteria is the use and degradation of 'dead' organic matter by the (heterotrophic) saprotrophs.

10.3.2 The nitrogen cycle

Nitrogen is a component of proteins and nucleic acids—and is therefore essential to all organisms. Gaseous nitrogen (dinitrogen) makes up over three-quarters of the Earth's atmosphere, but most organisms cannot use this form of nitrogen.

Many bacteria assimilate nitrogen as ammonia, primarily by incorporating it in the amino groups of amino acids. Some can also use nitrate, though the nitrate is first reduced to ammonia by *assimilatory nitrate reduction* (Fig. 10.2); this differs from *nitrate respiration* (section 5.1.1.2; Fig. 10.2) e.g. in that it does not yield energy.

Denitrification is an energy-yielding process (section 5.1.1.2) in which nitrate, or nitrite, is reduced to gaseous products, mainly dinitrogen and/or nitrous oxide; the process can be important in agriculture since it causes the loss of biologically useful nitrogen from the soil. *Nitrification* (Fig. 10.2) is an energy-yielding process which involves the oxidation of ammonia, or nitrite, by certain chemolithotrophs (section 5.1.2).

10.3.2.1   Nitrogen fixation

The 'fixation' of atmospheric nitrogen (reduction of nitrogen to ammonia) is apparently carried out only by certain types of bacteria (*diazotrophs*); diazotrophs include some cyanobacteria (e.g. *Anabaena, Nostoc*), some species of *Bacillus* and *Clostridium, Klebsiella pneumoniae* (some strains), and members of the families Azotobacteriaceae and Rhizobiaceae and of the order Rhodospirillales.

Fixation is catalysed by the enzyme complex *nitrogenase*. Electrons from a source such as hydrogen, or NADPH, are transferred e.g. to a ferredoxin (section 5.1.1.2) and thence to nitrogenase; the reduction of nitrogen requires much energy—about 12–16 molecules of ATP for each molecule of nitrogen fixed.

The ammonia produced by nitrogen fixation may be assimilated e.g. by the amination of 2-oxoglutarate to L-glutamate, or of glutamate to glutamine.

Nitrogenase is highly sensitive to oxygen. Many diazotrophs (e.g. clostridia) fix nitrogen anaerobically or microaerobically, and in some cyanobacteria the process is carried out within *heterocysts* (section 4.4.2); aerobic nitrogen-fixers have special mechanisms for protecting their nitrogenase.

Diazotrophs occur as free-living organisms in soil and water, and some are involved in symbioses (section 10.2.4.1).

10.3.2.2   Agriculture and the nitrogen cycle

Our knowledge of the roles of bacteria in the nitrogen cycle can be put to good use in improving agricultural food production. Food crops are often limited in yield by a shortage of available nitrogen in the soil; hence, by

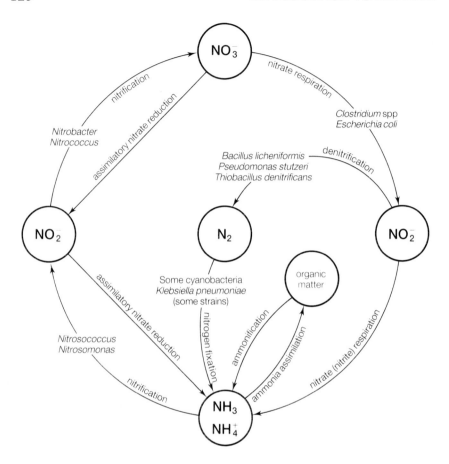

Fig. 10.2 The nitrogen cycle: some interconversions carried out by bacteria (section 10.3.2). Nitrification is an aerobic process, while nitrate respiration/denitrification and nitrogen fixation are typically anaerobic or microaerobic processes; in some cyanobacteria nitrogen fixation occurs in heterocysts (section 4.4.2). Nitrate, or ammonia, can be assimilated (used as a source of nitrogen) by many types of bacteria: the nitrate is reduced, intracellularly, to ammonia (*assimilatory nitrate reduction*). Ammonification is part of the mineralization process (section 10.3.4).

knowing how nitrogen is lost—and by exploiting nitrogen fixation—we can often take appropriate measures to increase crop yields.

Nitrogen is taken from the soil when crops are grown and harvested, and it may also be lost by denitrification and nitrification. Denitrification typically occurs under anaerobic conditions in the presence of nitrate and organic nutrients—conditions found e.g. in waterlogged farm soils; thus, denitrification can often be reduced by improving soil structure and drainage so as to minimize the development of anaerobic conditions.

The harmful effect of denitrification is obvious, but why does *nitrification* lead to a loss of nitrogen? The answer is that, although nitrate and ammonia are both soluble, ammonium ions adsorb readily to soil particles (clay particles typically bear a net negative charge) whereas nitrate ions do not; for this reason, nitrate is much more readily washed (*leached*) from the soil by rain or flooding. Hence, if nitrogen fertilizers are required it is better to choose ammonium compounds rather than nitrates. Nitrification can often be prevented by adding a 'nitrification inhibitor' to the fertilizer; such compounds primarily block the oxidation of ammonia, and one of them, *etridiazole*, is additionally useful as a fungicide—being used e.g. for soil and seed treatment against certain 'damping off' diseases.

Nitrogenous fertilizers can replace lost nitrogen, but they are expensive and can be afforded least by countries in greatest need of them. However, there are alternatives. For example, 'nitrogen-fixing plants' such as clover and lucerne can be included in crop rotation schemes, while plants such as *Azolla* (section 10.2.4.1) can be used as 'green manure'—a practice common in South-East Asia; in rice paddies, fertility can be increased by encouraging the growth of free-living nitrogen-fixing cyanobacteria. Even better would be the creation, by genetic manipulation, of plants capable of fixing nitrogen without help from prokaryotes.

10.3.3   The sulphur cycle

Sulphur is a component e.g. of the amino acids cysteine and methionine, of ferredoxins (section 5.1.1.2), and of cofactors such as coenzyme A. Green plants (and many bacteria) can assimilate sulphur in the form of sulphate, a substance commonly available in adequate amounts under natural conditions. Before incorporation, sulphate must be reduced to sulphide by *assimilatory sulphate reduction* (Fig. 10.3); this process differs from *sulphate respiration* (section 5.1.1.2; Fig. 10.3) in much the same way as assimilatory nitrate reduction differs from nitrate respiration (section 10.3.2). Some bacteria can assimilate sulphide direct from the environment.

In some habitats (e.g. stagnant anaerobic ponds) much sulphide is formed by sulphate respiration; this sulphide may, in turn, be used as electron donor (i.e. oxidized to sulphite or sulphate) by anaerobic photosynthetic bacteria (section 5.2.1.2).

Elemental sulphur can be used by some bacteria—see e.g. *Sulfolobus* in the Appendix.

10.3.4   Mineralization

Figures 10.1, 10.2 and 10.3 show that, in the cycles of matter, complex organic substances are broken down to simple inorganic materials such as carbon dioxide, sulphide, sulphate, ammonia and nitrate; this conversion of organic to inorganic matter is called *mineralization*.

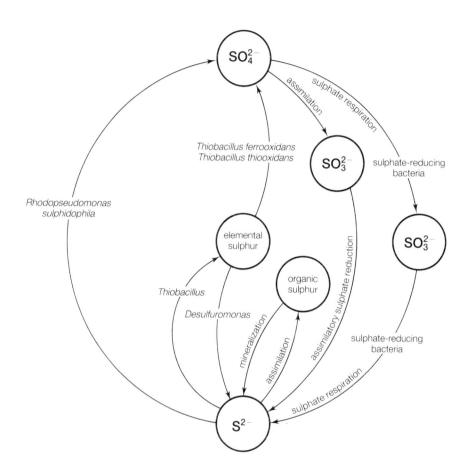

Fig. 10.3 The sulphur cycle: some interconventions carried out by bacteria. Many species can use sulphate (SO_4^{2-}) as a source of sulphur (needed e.g. for the synthesis of certain amino acids); this is shown in the figure as assimilatory sulphate reduction. In this process, sulphate is reduced, intracellularly, to sulphide, the sulphide being incorporated in different ways by different organisms; in e.g. *Escherichia coli*, sulphide is incorporated into *O*-acetylserine to form cysteine. Sulphate respiration (= dissimilatory sulphate reduction) is carried out e.g. by the 'sulphate-reducing bacteria'—organisms which use sulphate (or, e.g. sulphite) as a terminal electron acceptor in anaerobic respiration (section 5.1.1.2). *Desulfuromonas* uses elemental sulphur as a terminal electron acceptor in anaerobic respiration. *Thiobacillus* spp typically carry out aerobic respiration in which they oxidize e.g. sulphide (S^{2-}) and/or elemental sulphur. *Rhodopseudomonas sulfidophila* is one of a number of species which use sulphide as an electron donor in anaerobic phototrophic metabolism (section 5.2.1.2).

10.4   ICE-NUCLEATION BACTERIA

At temperatures just below 0 °C, some bacteria promote water-to-ice transition by acting as nuclei around which ice crystals can form. Some of these 'ice-nucleation bacteria' are commonly found on the surfaces of plants, and they have been implicated as contributory factors in frost damage in various agricultural crops; they include strains of *Erwinia, Pseudomonas* and *Xanthomonas*.

The ability to promote ice-crystal formation can be transferred e.g. to *E. coli* by transferring a particular fragment of DNA; this fragment appears to encode one of the proteins of the outer membrane.

10.5   BACTERIOLOGY *in situ*—FACT OR FICTION?

Ideally, every organism should be studied *in situ*, i.e. in its natural environment; biology is, after all, about the real, living world. Of course, some aspects—e.g. intracellular structures—have to be studied *in vitro* (in the laboratory), but, where possible, a cell's normal *behaviour* is best observed under conditions which most closely resemble its normal habitat.

Clearly, any meaningful *in situ* study demands an understanding of the particular environment since, without this, the design of the experiment can be faulty; unfortunately, many published studies which claim to be *'in situ'* involve obvious (sometimes extreme) distortions of nature, so that they are of little or no scientific value.

10.5.1   Membrane-filter chambers

To test for survival in rivers, suspensions of bacteria have been enclosed in 'membrane-filter chambers'—each essentially a wide plastic tube which is sealed at both ends by a membrane filter (pore size 0.22–0.45 μm); once a chamber had been immersed in a river, the bacteria inside it were considered to be in 'natural' conditions, i.e. the sample was believed to be 'in' the environment simply because the *chamber* had been immersed. In fact, in these experiments, the sample is largely cut off from the environment: it is shielded not only from light but also from microcurrents (and, hence, from fluctuations in temperature etc.); additionally, the normal dispersion and mixing of molecules (e.g. metabolic wastes) is severely restricted.

10.5.2   Conjugation *in situ?*

In an attempt to demonstrate conjugation (section 8.4.2) in a Welsh river and canal, suspensions of donors and recipients were mixed and then filtered through a membrane filter; the filter was placed (face-down) on a flat, sterile stone and held in place by a glass-fibre filter which was secured (to the stone) by rubber bands. After 24 hours immersion in the river or canal the whole was tested for transconjugants. The idea was to test for conjugation in simulated *epilithon* (the mixed community of organisms embedded in slime on the surfaces of underwater stones) [Bale, Fry & Day (1987) JGM *133* 3099–3107].

Was this a valid experiment? In real epilithon, conjugation would involve a chance contact between donor and recipient, the probability of contact depending mainly on the *numbers* of donors and recipients. Simulated epilithon should therefore contain comparable numbers of donors and recipients; higher numbers would invalidate the experiment by artificially increasing the probability of donor–recipient contact. Logically, one would first count (or estimate) potential conjugants in real epilithon and then use similar numbers in the experiment; however, such observations were apparently not made, the authors being seemingly unaware of the need to match reality in this vital aspect of the experiment. In the experiment, the filter was packed with 10^7 donors and 10^8 recipients/cm^2; according to the authors' own figures, this was over 100 times greater than the total concentration of viable bacteria in natural epilithon—yet the experiment still claimed to "mimic nature". Furthermore, this impossibly high concentration of 'epilithic' cells consisted *solely* of de-repressed donors (section 8.4.2.2) intimately mixed with a 12-fold excess of recipients of guaranteed compatibility. Nature is not like this, and it is pointless to pretend that it is. Quite apart from cell numbers, no evidence was given that the all-important initiation of donor–recipient contact had not already occurred during mixing and filtration of the cells—i.e. even *before* the filter had been placed *in situ*. Such experiments may be fun, but they contribute nothing to science.

There is a good case for *in situ* studies. However, they demand an appreciation and respect for the complexities of nature, an acceptance of reality, and an avoidance of gross, meaningless distortion. Experimentation has to be conducted rationally, on a proper scientific basis. Moreover, the interpretation of any *in situ* study would be less than truthful if it failed to acknowledge the artificiality of the method.

11 Bacteria in medicine

11.1 BACTERIA AS PATHOGENS

Some diseases are due to 'errors' in the body's chemistry, but in many diseases symptoms result from the activities of certain microorganisms, or their product(s), on or within the body; any microorganism which (given suitable circumstances) can cause disease is called a *pathogen*. Among the many diseases of microbial origin, some are due to fungi, some to viruses, some to protozoa, and some to bacteria; a number of the latter diseases are described briefly at the end of the chapter.

In some diseases the link between disease and pathogen is highly specific: such a disease can be caused only by the appropriate species, or by particular strains of that species. Anthrax, for example, is caused only by certain strains of *Bacillus anthracis*; these strains contain plasmids (section 7.1) which encode (i) the anthrax toxin and (ii) a capsule which protects the pathogen. In other cases a disease may be due to any of several different causal agents; an example is gas gangrene: a disease which can be caused by one (or more) of several different species of *Clostridium*.

Sometimes a disease may be due to an organism which does not usually behave as a pathogen and which may actually be a member of the body's own microflora (Table 11.1); for example, species of *Bacteroides*, which are common in the intestine, can sometimes give rise to peritonitis following accidental or surgical trauma in the lower intestinal tract. Organisms such as these are called *opportunist pathogens*.

Disease does not *necessarily* follow exposure to a given 'causal agent'. In fact, the occurrence (or otherwise) of disease typically depends on various factors—including the degree of resistance of the host and the *virulence* (capacity to cause disease) of the pathogen.

11.2 THE ROUTES OF INFECTION

The skin is normally an effective barrier to pathogens, but skin may be broken—e.g. by wounding, surgery or the 'bites' of insects etc. Wounds may admit any of a variety of potential pathogens capable of causing systemic disease (disease affecting the entire body) or localized disease. Bacterial pathogens which can enter via 'bites' include the causal agent of bubonic plague.

Table 11.1   Human microflora: some of the bacteria commonly associated with the human body

Location	Species of
Colon	*Bacteroides*
	Clostridium
	Escherichia
	Proteus
Ear	*Corynebacterium*
	Mycobacterium
	Staphylococcus
Mouth	*Actinomyces*
	Bacteroides
	Campylobacter
	Streptococcus
Nasal passages	*Corynebacterium*
	Staphylococcus
Nasopharynx	*Streptococcus*
Skin	*Propionibacterium*
	Staphylococcus
	Others (according e.g. to personal hygiene and environment)
Urethra	*Acinetobacter*
	Escherichia
	Staphylococcus
Vagina (adult, pre-menopausal)	*Acinetobacter*
	Corynebacterium
	Lactobacillus
	Staphylococcus

Mucous membranes—such as those of the intestinal, respiratory and genitourinary tracts—tend to be more vulnerable than skin, and infections commonly begin at these sites. In pneumonia and whooping cough, for example, infection begins at the respiratory surfaces, while in cholera and typhoid it begins at the intestinal mucosa.

11.2.1   Adhesion as a factor in infection

In many diseases there is an early phase in which the pathogen adheres to particular sites in the host. The need for attachment becomes clear when we consider, for example, that the common sites of infection, the mucous membranes, are continually flushed by their own secretions and may be subject to movements such as peristalsis—factors which tend to discourage the establishment of a pathogen. Adhesion may also help a pathogen to compete more effectively with the host's own microflora.

Adhesion is essential for the virulence of *enterotoxigenic* strains of *E. coli* (so-called ETEC); these strains adhere to the duodenal mucosa and produce enterotoxin(s) responsible e.g. for many of the cases of travellers' diarrhoea. Adhesion in this case is due to specific types of fimbriae (section 2.2.14.2) which are encoded by plasmids (section 7.1).

Tooth decay (*dental caries*) is promoted by bacteria which adhere to tooth surfaces and gum margins and which contribute to *dental plaque*: a film composed mainly of bacteria, bacterial products, and salivary substances. *Streptococcus mutans*, a common component of plaque, forms extracellular water-insoluble glucans which assist bacterial adhesion; waste products of bacterial metabolism (e.g. lactic acid) cause localized demineralization in the teeth, permitting bacterial penetration.

Adhesion appears to be important also in the further decay of *filled* teeth, bacterial colonization occurring in the small gap between the filling material and the wall of the cavity [Buchmann *et al.* (1990) MEHD *3*, 51–57] (Plate 1, centre); from this site, bacteria may penetrate the dentinal tubules and bring about destruction of the dentine.

11.3 PATHOGENESIS: THE MECHANISM OF DISEASE DEVELOPMENT

How does a pathogen cause disease? Different pathogens act in different ways. Some produce toxins (or other substances) which disrupt specific physiological processes, while others invade particular cells or tissues (and may also form toxins). Even when localized in the body, such infections commonly have systemic effects. In some diseases, symptoms result from an 'over-reaction' of the body's own defence mechanisms. A few examples of pathogenesis are given below.

11.3.1 Toxin-mediated pathogenesis

Cholera involves e.g. vomiting and a profuse diarrhoea which eventually becomes virtually water. The pathogen (certain strains of *Vibrio cholerae*) multiplies in the gut and forms an enterotoxin which acts on mucosal cells in the small intestine. Within the mucosal cells, the toxin stimulates an enzyme, adenylate cyclase, causing an increase in cellular cAMP (Fig. 7.12); this, in turn, leads to a massive outflow of Na^+, Cl^- and water into the lumen of the intestine—and the dehydration characteristic of cholera.

Botulism involves muscle paralysis, and death may result e.g. from mechanical (muscular) failure of the respiratory system. The pathogen (strains of *Clostridium botulinum*) forms a toxin which binds to nerve–muscle junctions, inhibiting the release of acetylcholine and (hence) inhibiting nervous stimulation of the muscles. Disease can result from the ingestion of pre-formed toxin—usually in toxin-contaminated foods such as cooked meats, sausage, and improperly canned vegetables; that is, the pathogen *itself* need not be ingested.

Tetanus ('lockjaw') involves uncontrollable contractions of the skeletal muscles, often leading to death by asphyxia or exhaustion. The disease develops e.g. when deep, anaerobic wounds are contaminated with the pathogen, *Clostridium tetani*. *C. tetani* produces a toxin (*tetanospasmin*) which acts on certain cells (interneurones) in the central nervous system; by inhibiting the release of glycine from these cells, tetanospasmin permits the simultaneous contraction of both muscles in a protagonist–antagonist pair.

In some diseases (e.g. diptheria) the pathogen forms a *phage*-encoded toxin (section 9.4), while some other toxins (e.g. one formed by ETEC—section 11.2.1) are encoded by plasmids (section 7.1).

11.3.2 Pathogenesis involving other bacterial products

Aggressins are products which can promote the invasiveness of a pathogen. For example, certain bacteria, including most coagulase-positive staphylococci, produce *hyaluronate lyase* ('spreading factor'): an enzyme which cleaves hyaluronic acid (a component of the intercellular 'cement' in animal tissues); in at least some cases, this enzyme may assist bacterial penetration of an infected site.

Cystic fibrosis is an hereditary disease characterized e.g. by severe bronchial congestion. In some cases the respiratory tract becomes colonized by certain strains of *Pseudomonas aeruginosa* which produce a highly mucoid (alginate) slime that aids congestion and makes for a poor prognosis.

11.3.3 Pathogenesis involving destruction of host cells or tissues

Typhoid, caused by *Salmonella typhi*, involves e.g. intestinal symptoms and septicaemia. (*Septicaemia*: systemic illness resulting from bacteria in the blood.) After ingestion, the pathogen invades mucosa in the small intestine. In some cases, inflammation in the ileum is so intense that it causes local necrosis (death) of tissue and haemorrhage.

Enteroinvasive strains of *E. coli* (EIEC) can invade and destroy cells in the small intestine and colon, causing e.g. abdominal pain, profuse fluid stools, and dehydration. At least some of the strains are toxigenic (i.e. able to form toxin), and specific mechanisms for adhesion (section 11.2.1) seem to be involved.

Oroya fever, which occurs in parts of South America, involves e.g. fever and anaemia; the causal agent, *Bartonella bacilliformis*, is transmitted via the 'bites' of sandflies. *B. bacilliformis* grows in and on erythrocytes (red blood cells) and in the endothelial cells of the host; bacterial growth leads to the destruction of erythrocytes etc. and to associated symptoms.

11.3.4   Endotoxic shock

Each of the toxins mentioned in section 11.3.1 is an *exotoxin*: a protein toxin which is released by the toxigenic cell and which has a single, specific mode (and site) of action. By contrast, the *endotoxins* of Gram-negative bacteria are components of lipopolysaccharides (section 2.2.9.2)—primarily lipid A, though the polysaccharide component may be important in conferring water-solubility. Blood-borne endotoxins (on whole cells or fragments of lysed cells) seem to bring about various non-specific reactions, including the classical signs of shock (e.g. a fall in blood pressure); in general, the significance of endotoxins in disease has still not yet been decided.

11.4   THE BODY'S DEFENCES

11.4.1   Constitutive defences

Constitutive defences are non-specific defence mechanisms which are operative all the time.

To any potential pathogen, the normal healthy body presents a variety of obstacles and barriers. The skin, for example, is more than a simple physical barrier to infection. To most bacteria it is a hostile environment: water is scarce, and sites are occupied by the well-adapted skin microflora, some of which produce antibacterial fatty acids from lipids secreted by the sebaceous glands.

Mucous membranes, too, have their own defences: the secretions which bathe these tissues actively discourage the establishment of a pathogen, both by their mechanical flushing action and by their antibacterial substances; tears, for example, contain *lysozyme* (Fig. 2.7, caption). Then there is the resident microflora with which any would-be pathogen must compete if it is to become established.

If a pathogen penetrates the outer defences it is immediately faced with the inner defences. Within the tissues and circulatory systems, certain specialized cells (*phagocytes*) engulf and destroy particles of 'foreign' matter—including many types of microorganism; these scavenging cells (which include e.g. *macrophages*) can usually prevent the establishment of a pathogen in the (nutrient-rich) tissues of the body.

If the pathogen persists it may cause *inflammation*: reddening, swelling, warmth and pain at the affected site. Inflammation can be caused by various agents—including e.g. heat and chemical irritants as well as micro-organisms—and although non-specific, some of its effects can inhibit a pathogen. For example, inflammation involves an increased outflow of plasma (and of certain types of cell) from small blood vessels into the affected tissues (which therefore swell); the pathogen is therefore exposed to increased amounts of certain antimicrobial factors which occur in normal plasma. Inflammation is thus an important part of the body's generalized reaction to pathogens.

11.4.1.1 Complement

Normal plasma contains *complement*: a system of individual proteins which, in the presence of certain types of molecule, undergo a 'cascade' of reactions in which each component of the system is activated in turn; on activation, various components of the system carry out specific physiological functions.

Complement can be activated e.g. by lipopolysaccharides (section 2.2.9.2). (In this case, the cascade of reactions follows the so-called *alternative pathway* of complement activation.) Activation by a Gram-negative bacterium has several consequences. Various components of complement bind sequentially to one another at the cell surface; when the last components have bound, the resulting *membrane attack complex* (consisting of components C5b, 6, 7, 8 and 9) produces a hole in the outer membrane. This can lead to cell lysis (*immune cytolysis*). (Complement is thus an early, rapid form of non-specific defence.) Meanwhile, other components may carry out different functions. For example, component C3a can cause release of *histamine* from mast cells; histamine causes e.g. increased permeability in certain small blood vessels— thus contributing to the inflammatory response.

11.4.2 The adaptive response

In addition to constitutive defences, the body can also respond *specifically* to a given pathogen. An individual cell can be recognized by its 'chemical fingerprint'—i.e. the molecules (such as lipopolysaccharides) which charac-terize the cell. Such molecules may act as *antigens*, i.e. their presence in the body may stimulate certain white blood cells (particular 'strains' of B lymphocytes) to secrete proteins called *antibodies*; an antibody can combine specifically with the antigen that induced it. (In an adult, the B lymphocytes consist of some one million 'strains', each strain being able to recognize a particular type of antigen and to secrete the 'matching' antibody.)

Antibodies induced by a given pathogen can combine with that pathogen, but of what use is this? Suppose that antibodies bind to cell-surface molecules (such as lipopolysaccharides). Most antigen–antibody complexes will automatically activate the complement system (section 11.4.1.1) and bind components of that system—the sequence of binding in this case following the so-called *classical pathway*; one component (C3b) confers on the antigen–antibody–complement complex the ability to bind strongly to phagocytes (*immune adherence*) so that the whole becomes highly susceptible to phagocytosis. (This phenomenon, in which antigens/cells are made more susceptible to phagocytosis, is called *opsonization*.)

Following exposure to a given antigen, some B lymphocytes of the appropriate strain secrete antibodies, but others become *memory cells*. On subsequent exposure to the same antigen, the memory cells enable the body to respond with a more rapid and vigorous production of specific antibodies (the *secondary response*).

Just as components of a pathogen can act as antigens, so too can many bacterial products, including toxins. The combination of a toxin with its antibody abolishes its harmful properties and assists in its elimination from the body. However, in certain toxin-mediated diseases (e.g. tetanus) death can occur before the body has had time for an effective antibody response.

The production of antibodies can be stimulated artificially by *vaccination*, the object of which is to help the body defend itself in the event of an attack by a given pathogen. In vaccination, the *vaccine*—containing antigens of the pathogen (typically killed cells, or cell components)—is introduced into the body by injection or, sometimes, orally. The body responds by producing the corresponding antibodies; additionally, it becomes 'primed', i.e. the corresponding memory cells (see above) can respond, rapidly, with a vigorous production of specific antibodies on any subsequent exposure to the pathogen.

Vaccination against a toxin (such as tetanospasmin—section 11.3.1) involves the administration of a modified form of the toxin (a *toxoid*) which is no longer harmful but which has kept its specific antigenic characteristics; a toxoid thus induces the formation of anti-toxin antibodies.

Vaccination is typically given *prophylactically*, i.e. with the object of preventing disease. In e.g. botulism (section 11.3.1) an *antiserum* (in this case, a serum containing pre-formed anti-toxin antibodies) is generally used in the treatment of the disease.

Another type of adaptive defence is mediated by cells rather than by antibodies: so-called *cell-mediated immunity* (CMI). CMI has many aspects, some being more applicable e.g. to anti-cancer mechanisms than to antibacterial defence. However, certain T lymphocytes can react, specifically, to bacterial antigens—contact with a specific antigen causing the T cell to release certain substances (*cytokines*, formerly called *lymphokines*) which can attract and 'activate' particular types of phagocyte. On activation, a macrophage becomes more aggressive towards engulfed bacteria, and it also secretes increased amounts of antibacterial substances (such as hydrogen peroxide); it can kill certain bacteria (e.g. some species of *Mycobacterium*) which are not normally killed in non-activated phagocytes. (*Legionella* spp. can inhibit phagocyte activation [Dowling *et al.* (1992) MR 56 32–60].) Nevertheless, in some diseases CMI appears to contribute to the symptoms; in tuberculosis, for example, at least some of the tissue damage which occurs at the *tubercles* (localized sites of infection) appears to be due to the leucocytes (white blood cells) which concentrate at these sites.

11.5   THE PATHOGEN: VIRULENCE FACTORS

Actively aggressive products such as toxins and aggressins (sections 11.3.1 and 11.3.2) are clearly *virulence factors*. However, so, too, are those products and strategies which help a pathogen to evade host defences. Some of these other factors are considered below. (A recent minireview discusses the control of bacterial virulence factors [Mekalanos (1992) JB 174 1–7].)

11.5.1   Capsular camouflage

Some pathogens have anti-phagocytic capsules. For example, in certain strains of *Streptococcus pyogenes* the capsule is made of hyaluronic acid, a component of animal tissues; this 'camouflage' seems to give some protection against phagocytosis. The poly-D-glutamic acid capsule of *Bacillus anthracis* (causal agent of anthrax) appears to have a similar role.

11.5.2   Leucocidins

Substances which can damage or kill phagocytes (*leucocidins*) are secreted by certain pathogenic staphylococci and streptococci. The staphylococcal *Panton–Valentine leucocidin* specifically lyses macrophages and polymorphonuclear leucocytes.

11.5.3   Antigenic variation

Certain pathogens can change their cell-surface chemistry—and, hence, their antigens; this may help, even temporarily, to avoid the effects of specific antibodies. For example, in relapsing fever (caused by species of *Borrelia*) there are several cycles of fever and remission, and bacteria isolated from the blood during each period of fever are found to have different surface antigens. In *B. hermsii*, antigens involved in antigenic variation are encoded by a linear (i.e. not ccc) multicopy plasmid (section 7.3.1); antigen switching appears to involve site-specific recombination (section 8.2.2) between different individual plasmids in a given cell. SSR is also involved in the mechanism for phase variation in *Salmonella* (Fig. 8.3c).

11.6   THE TRANSMISSION OF DISEASE

In some diseases the pathogen does not normally spread from one individual to another. Examples include e.g. tetanus and gas gangrene; in these diseases the pathogen is usually a wound contaminant which typically originates in the soil rather than in another, infected individual.

Other diseases spread from one person (or animal) to another, either directly or indirectly. In relatively few (bacterial) diseases does transmission require direct contact between an infected individual and a healthy one, and these diseases generally involve pathogens which cannot survive for long periods of time outside the body; an example is the venereal disease syphilis, caused by *Treponema pallidum*.

Most diseases spread indirectly from person to person, usually in a way related to the normal route of infection (section 11.2). For example, pathogens which infect via the intestine are commonly transmitted in contaminated food or water: gastroenteritis, dysentery, typhoid and cholera are usually spread in this way. Such transmission generally involves some connection between the food or water and the faeces of a patient who is suffering from the disease; contamination may occur, for example, when the hands of a food-handler carry traces of faecal matter, when a housefly lands alternately on faeces and food, or when sewage has leaked into a source of drinking water. Often the pathogen can be traced back—via the food or water—to individual(s) suffering from the disease. Sometimes, however, the source of a pathogen is a person who is *not* suffering from the disease but who is nevertheless playing host to the pathogen and acting as a *reservoir* of infection; such an apparently healthy individual who is a source of pathogenic organisms is called a *carrier*. One notorious carrier, a cook by the name of Mary Mallon, transmitted typhoid fever to nearly thirty people before she was traced, and the name 'typhoid Mary' is sometimes used to refer to an actual or suspected carrier in outbreaks of typhoid and other diseases.

Pathogens which infect via the respiratory tract are often transmitted by so-called *droplet infection*. When a person coughs or sneezes, or even speaks loudly, minute droplets of the mucosal secretions are expelled from the mouth; these droplets can contain pathogens if such pathogens are present on the respiratory surfaces. Since the smallest droplets can remain suspended in air for some time, they can be inhaled by other individuals and can therefore act as vehicles for the transmission of pathogens. Diphtheria, whooping cough and pulmonary tuberculosis are examples of diseases which can be transmitted in this way.

A few bacterial pathogens are transmitted from one person to another by a third organism called a *vector*. For example, bubonic plague (caused by *Yersinia pestis*) is transmitted by fleas, while epidemic typhus (caused by *Rickettsia prowazekii*) is typically transmitted by lice; tularaemia (caused by *Francisella tularensis*) can be transmitted e.g. by biting flies or by ticks. Oroya fever (section 11.3.3) is another example.

11.7 LABORATORY DETECTION AND EXAMINATION OF PATHOGENS

11.7.1 Culture

Commonly, attempts are made (where possible) to culture the pathogen from a sample of e.g. sputum, pus, urine or faeces; before culture, the urine may be centrifuged (and the sediment used as an inoculum), while a faecal sample may be dispersed e.g. in Ringer's solution. The choice of medium, and the conditions of incubation, will depend on the inoculum and on the nature of the suspected pathogen; where appropriate, an enrichment medium is used (section 14.2.1).

11.7.1.1 Blood culture

Blood culture is used e.g. to detect septicaemia (in diseases such as typhoid). Essentially, 5–10 ml of the patient's blood (taken aseptically—section 14.3) is added to 50–100 ml of a medium such as trypticase soy broth; the medium usually contains a blood anticoagulant, and may also contain e.g. particular enzymes to inactivate antibiotic(s) carried over in the blood. The inoculated medium is incubated, and is examined, daily, by subculture to an appropriate solid medium.

11.7.2 Immunofluorescence microscopy

This technique permits detection of a specific type of organism in a smear (section 14.9) containing a mixture of organisms. Antibodies (section 11.4.2), specific to the given organism, are first linked chemically to a fluorescent dye (e.g. fluorescein); the *conjugate* (i.e. suspension of dye-linked antibodies) is added to the smear, unbound conjugate is rinsed off, and the slide is examined by *epifluorescence microscopy*. In epifluorescence microscopy, ultraviolet radiation is beamed onto the specimen; visible light from any bound, fluorescent antibodies is seen, in the usual way, via the objective lens of the microscope.

11.7.3 Complement-fixation tests (CFTs)

CFTs can detect specific antigens or antibodies; they depend on the binding ('fixation') of complement (section 11.4.1.1) by most types of antigen-antibody complex. If, for example, antigen is added to a sample of patient's serum, fixation of complement in the sample would typically indicate the presence of specific ('matching') antibodies in that serum. In practice, the serum is first heated (56 °C/30 min) to destroy the patient's own complement, and serial dilutions are prepared; then, a known, standard amount of complement is added to each dilution. To detect antibodies, a standard amount of specific antigen is added to each dilution, and the whole is incubated e.g. for 18 hours at 4 °C. In a given dilution of serum, the amount of free (unbound) complement *remaining* after incubation indicates the presence/quantity of specific antibodies in that dilution. The residual complement is assayed by adding a *haemolytic system* (a suspension of 'sensitized' erythrocytes) to each dilution and noting the proportion of erythrocytes which undergo complement-mediated lysis. If, for example, maximum lysis occurs in the highest concentration of serum, the test is *negative*, i.e. the serum contains maximum residual complement (no detectable antibodies).

A CFT is used e.g. in the Wassermann test for syphilis.

11.7.4 Enzyme-linked immunosorbent assay (ELISA)

ELISA is a highly sensitive test for specific antigens or antibodies. The *principle* is shown in Fig. 11.1.

11.7.5 Characterization of the pathogen

Once obtained in pure culture, the pathogen is usually examined by the procedures and tests outlined in Chapter 16; this typically permits identification to the level of species or serotype. Commonly, a pathogen's pattern of resistance to a range of antibiotics is determined—often by a disc diffusion test (section 15.4.11.1).

11.8 PREVENTION AND CONTROL OF TRANSMISSIBLE DISEASES

Once we know how a disease spreads, it's often possible to devise methods for preventing or limiting its spread. Clearly, any disease which spreads only by direct physical contact can be prevented simply by avoiding such contact with infected persons. For other transmissible diseases, prevention or control may involve blocking the route of infection from one person to another. Thus, the spread of diseases such as typhoid and cholera can be halted by measures such as: (i) improvement in personal hygiene—e.g. washing hands after visiting the lavatory; (ii) protection of food etc. from flies and other insects likely to carry pathogens, and reduction in the numbers of such insects by the use of insecticides; (iii) protection of drinking

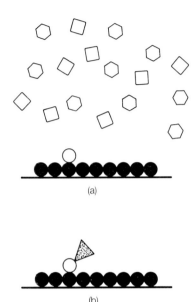

Fig. 11.1   The *principle* of enzyme-linked immunosorbent assay (ELISA). Here, an attempt is being made to detect a low concentration of a particular antibody (open circle) among high concentrations of other antibodies (open squares, hexagons) in a sample of serum.

(a)   Specific antigen (solid circles) is immobilized e.g. on the inner surface of a plastic test-tube, and is exposed to the serum. The single molecule of specific antibody binds to its corresponding antigen. Other antibodies, which do not bind, are subsequently removed by washing.

(b)   The single, bound antibody is then detected by means of anti-immunoglobulin antibodies—i.e. antibodies whose corresponding antigens are themselves antibodies, and which thus bind specifically to other antibodies; before use, each anti-immunoglobulin antibody is chemically linked to a particular enzyme. An enzyme–anti-immunoglobulin antibody complex (stippled triangle) has bound to the solitary bound antibody, so that the latter is now labelled with an enzyme; by detecting the enzyme one can now detect the antibody. The enzyme is detected by adding a suitable substrate which, in the presence of the enzyme, yields a measurable product which is *amplified* by continued enzymic action.

water from contamination by sewage, and the effective treatment of communal water supplies with antimicrobial agents such as chlorine; (iv) disinfection of small quantities of untreated water before consumption— e.g. by boiling or by treatment with a disinfecting agent such as halozone; (v) detection and isolation of carrier(s) (section 11.6), if involved.

Diseases which spread by droplet infection (section 11.6) are generally more difficult to deal with. The physical exclusion of droplets (e.g. with face masks) is usually impracticable, and one of the main control measures in diseases such as diphtheria and whooping cough involves the protection of susceptible individuals by vaccination (section 11.4.2). In such diseases there is also some advantage in effectively isolating (*quarantining*) sick individuals until they are no longer able to act as sources of the pathogen.

Diseases spread by vectors can be controlled by eliminating the vector or by reducing its numbers. Such control is applicable e.g. in typhus and bubonic plague.

11.9   A NOTE ON THE TREATMENT OF BACTERIAL DISEASES

The availability of a wide range of antibiotics (section 15.4) means that most bacterial diseases can be treated by *chemotherapy*, i.e., therapy involving the use of *chem*ical agents (such as antibiotics). ('Keemotherapy' presumably involves the use of *keem*ical agents!) Considerations involved in the use of a given antibiotic include: (i) effectiveness against the pathogen; (ii) occurrence of resistant strains of the pathogen; (iii) antagonism with other forms of treatment (if any); (iv) possible side-effects.

In some diseases (e.g. botulism, diphtheria) treatment commonly includes administration of *antitoxin*, i.e. a preparation containing antibodies to the relevant toxin.

In gas gangrene, treatment includes surgical removal of dead/infected tissue, and may include the use of hyperbaric oxygen.

11.10   SOME BACTERIAL DISEASES

The following note-form descriptions are intended simply to give some idea of the range of types of infection which can involve bacteria. A few of the diseases (e.g. conjunctivitis) can be caused by agents other than bacteria, but most of the diseases listed are caused only by bacteria—and often only by specific pathogens.

The name of the disease is followed by (i) the name(s) of the causal agent(s), (ii) the characteristic symptoms, and (iii) the route(s) of infection, where known.

Anthrax   *Bacillus anthracis* (virulent strains). Localized pustule on skin (anthrax boil) or, rarely, lung infection (woolsorters' disease) or intestinal anthrax; septicaemia can occur in untreated cases. Infection via skin wounds, inhalation, ingestion.

Bacterial vaginosis   Disease associated with increased numbers of e.g. species of *Bacteroides* and *Gardnerella*. Malodorous discharge. Vagina typically less acidic and more 'reducing' than is normal. 'Clue cells' (vaginal epithelial cells coated with small Gram-negative rods) typical in smears.

Botulism   *Clostridium botulinum* (virulent strains). Generalized weakness, defective vision, respiratory paralysis (section 11.3.1). In adults: commonly, ingestion of pre-formed toxin; less commonly(?): intestinal infection with toxigenic *C. botulinum* [Sonnabend *et al.* (1987) Lancet *i* 357—360]. Rarely, wound contamination with *C. botulinum*. In infants: a 'toxicoinfection' in which *C. botulinum* produces toxin in the intestinal tract.

Cellulitis   Usually, strains of *Staphylococcus* or *Streptococcus*. Diffuse, spreading inflammation typically affecting subcutaneous tissues. Infection via wounds etc.

Cholera   *Vibrio cholerae* (virulent strains). Intestinal infection. Copious watery stools (section 11.3.1) and dehydration. Infection via oral–faecal route, usually via contaminated water.

Conjunctivitis   Various, e.g. *Haemophilus influenzae*. Inflammation of the conjunctivae (mucous membranes of the eye).

Cystitis   Various, e.g. *Escherichia coli* (strains which have 'P fimbriae' adhere to uroepithelium), *Proteus* spp. Inflammation of the urinary bladder. Infection: downwards from the kidneys or (more commonly) upwards from the urethra, with *E. coli* and *Proteus* spp being common causal agents in the latter type.

Diphtheria   *Corynebacterium diphtheriae* (virulent strains). Fever. Membrane forms usually in the tonsillar and/or adjacent regions, inhibiting breathing and/or swallowing. Toxin may cause e.g. myocarditis. A carrier state (section 11.6) is recognized. Droplet infection (section 11.6) or ingestion of contaminated food, milk etc.

Dysentery (bacillary)   *Shigella* spp. Intestinal inflammation, pain. Copious fluid stools often with blood, mucus, pus. Dehydration. Infection via oral–faecal route, e.g. contaminated food, water.

Erysipelas   *Streptococcus pyogenes*. A form of cellulitis, often affecting the face; fever, prostration, septicaemia may occur. Infection via wounds etc.

Food poisoning   Various, e.g. *Bacillus cereus*, *Campylobacter jejuni*, *Clostridium perfringens*, *Escherichia coli*, *Salmonella typhimurium*, *Staphylococcus aureus*, *Vibrio parahaemolyticus*, *Yersinia enterocolitica*. Acute gastroenteritis. Abdominal discomfort/pain. Usually diarrhoea (but may be little/none in staphylococcal food poisoning and in one type caused by *B. cereus*). Nausea and vomiting are common, but not invariably present. Oral–faecal route; ingestion of pathogen and/or toxin in contaminated food.

Gas gangrene   Typically, *Clostridium perfringens* (type A), *C. septicum* and/or *C. novyi*. Spreading necrosis (death) which may start in any of various tissues. Tissues swell and contain pockets of gas (formed by bacterial metabolism). Infection via wounds, boils etc.

Gonorrhoea   *Neisseria gonorrhoeae* ('gonococcus'). Discharge from the genitourinary tract; other symptoms may occur. Sexual contact.

Legionnaires' disease   *Legionella pneumophila*. Malaise, myalgia, fever, pneumonia. Infection presumed to occur via contaminated aerosols.

Leptospirosis *Leptospira interrogans* (a spirochaete). Mild form: fever, headache, myalgia; typically impaired kidney function. Severe form (*Weil's disease, infectious jaundice*): above symptoms with e.g. vomiting, diarrhoea, liver enlargement, jaundice, haemorrhages, meningitis. Infection via wounds or mucous membranes; the pathogen occurs e.g. in water contaminated with urine from infected animals.

Leprosy *Mycobacterium leprae*. According to the type of leprosy, lesions in any of a variety of tissues (commonly including the skin) and/or destruction of nerves in the peripheral nervous system.

Meningitis Various, e.g. *Neisseria meningitidis, Haemophilus influenzae* type b (common cause in infants and children), *Streptococcus pneumoniae*. Inflammation of the meninges (membranes which cover the brain and spinal cord). Droplet infection, head wounds etc.

Plague (bubonic) *Yersinia pestis*. Fever, haemorrhages, *buboes* (inflamed, swollen, necrotic lymph nodes); septicaemia may occur and lead to e.g. meningitis. Infection via the 'bite' of a flea.

Pneumonia Various, e.g. *Streptococcus pneumoniae, Haemophilus influenzae* type b, *Staphylococcus aureus, Klebsiella pneumoniae*. Fever, difficulty in breathing, chest pain, cough. Droplet infection etc.

Q fever *Coxiella burnetii*. Headaches, fever, muscular pain and (often) respiratory symptoms. Inhalation, or e.g. ingestion of contaminated milk.

Scarlet fever *Streptococcus pyogenes*. Commonly in children: sore throat, fever, swelling of cervical lymph nodes, rash. Droplet infection, ingestion of contaminated milk etc.

Syphilis *Treponema pallidum*. A lesion (*chancre*) at the site of infection (typically genital mucosa) followed by a skin rash and, after months/years in untreated cases, lesions in e.g. heart, central nervous system etc. Infection by direct, particularly sexual, contact.

Tetanus *Clostridium tetani* (virulent strains). Sustained, involuntary muscular contraction (section 11.3.1). Typically, wound contamination.

Trachoma *Chlamydia trachomatis*. Conjunctivitis, scarring of conjunctival tissue, inturned eyelids, abrasion of the cornea by the eyelashes, causing ulceration. Infection occurs contaminatively.

Tuberculosis (pulmonary) Usually, *Mycobacterium tuberculosis*. Lesions develop in the lungs, become encapsulated, and subsequently break down— allowing dissemination of the pathogen. Infection commonly by inhalation.

Tularaemia *Francisella tularensis*. Chills, fever, nausea, headache, vomiting, prostration; sometimes severe gastrointestinal symptoms/septicaemia. Contaminative infection via wounds or mucous membranes, or by the 'bite' of a tick or biting fly etc.

Typhoid *Salmonella typhi*. Fever, transient rash, intestinal inflammation, septicaemia, sometimes with tissue necrosis and intestinal haemorrhage. A carrier state (section 11.6) is recognized, *S. typhi* typically localizing in the gall bladder. Oral-faecal route, e.g. contaminated food or water.

Typhus (classical, epidemic) *Rickettsia prowazekii*. Headaches, sustained fever and a rash which may haemorrhage, muscular pain. Contamination of a wound or 'bite' with louse faeces containing *R. prowazekii*; possibly, inhalation of dried, infected louse faeces.

Urethritis Various, e.g. *Neisseria gonorrhoeae* (in gonorrhoea), or, in non-gonococcal (= 'non-specific') urethritis, *Chlamydia trachomatis*, *Mycoplasma hominis*. Inflammation of the urethra. Commonly, infection involves sexual contact.

Whooping cough *Bordetella pertussis*. Paroxysms of coughing, each followed by an inspiratory 'whoop'. Droplet infection.

12 Applied bacteriology I: food

12.1 BACTERIA IN THE FOOD INDUSTRY

Bacteria are used for: (i) fermentations in the dairy industry; (ii) the processing of raw materials in the manufacture of coffee and cocoa; (iii) the manufacture of food additives; (iv) other processes, such as vinegar production. (Bacterial involvement in the manufacture of food for farm animals is considered in Chapter 13.)

12.1.1 Dairy products

The manufacture of butter, cheese and yoghurt involves a *homolactic fermentation* (section 5.1.1.1) in which the lactose in milk is metabolized to lactic acid.

Butter ('cultured creamery butter') is usually made from pasteurized cream to which a *lactic acid starter* culture has been added; the starter contains e.g. *Streptococcus cremoris* and/or *S. lactis* as the main lactic acid producers, together with *S. lactis* subsp. *diacetylactis* and/or *Leuconostoc cremoris* as the main contributors of diacetyl. Lactic acid and diacetyl contribute to the flavour, diacetyl giving the characteristic 'buttery' odour and taste. 'Sweet cream butters' are made without a starter.

Cheese is made by the coagulation and fermentation of milk, the different types of cheese reflecting differences e.g. in the source of milk (cows', goats' milk etc.), and in the types of microorganism used—commonly species of *Streptococcus* and *Lactobacillus*. Fermentation lowers the pH, thus helping in the initial coagulation of milk protein; additionally, minor products of fermentation (e.g. acetic and propionic acids) give characteristic flavours. In Swiss cheeses such as Emmentaler and Gruyère, the typical flavour of propionic acid is due to the use of *Propionibacterium* spp.

Yoghurt is usually made from pasteurized low-fat milk which is high in milk solids. The milk is inoculated with *Lactobacillus bulgaricus* and *Streptococcus thermophilus* and incubated at 35–45 °C for several hours; the pH falls to about 4.3, coagulating the milk proteins. The bacteria act co-operatively: *L. bulgaricus* breaks down proteins to amino acids and peptides—which stimulate the growth of *S. thermophilus*; formic acid produced by *S. thermophilus* stimulates the growth of *L. bulgaricus*, which forms most of the lactic acid.

12.1.2 Coffee and cocoa

The manufacture of coffee from ripe coffee fruits requires the initial removal of a sticky mucilaginous mesocarp from around the two beans in each fruit. The outer skin of the fruit is disrupted, and the whole is left to ferment. The mucilage is degraded both by the fruit's own enzymes and by microbial extracellular enzymes. As well as e.g. yeasts (single-celled fungi), the important organisms in this process include pectinolytic species of e.g. *Bacillus* and *Erwinia*. After fermentation, the beans are washed, dried, blended and roasted.

Cocoa is made from the seeds (beans) of the cacao plant, the fruit of which is a pod containing up to 50 beans covered in a white mucilage. The mucilage is fermented by yeasts, which produces ethanol, and the ethanol is oxidized to acetic acid by certain aerobic bacteria; other organisms are also present. Once free of mucilage, the beans—which darken during the week-long fermentation—are dried and roasted.

12.1.3 Food additives

Monosodium glutamate, the ubiquitous 'flavour enhancer', is manufactured from L-glutamic acid; this latter product is obtained commercially from certain bacteria—e.g. strains of *Corynebacterium glutamicum* grown aerobically on substrates such as molasses or hydrolysed starch. These organisms form glutamic acid from 2-oxoglutaric acid; because they lack the appropriate enzyme, commercial strains of *C. glutamicum* cannot metabolize 2-oxo-glutaric acid in the TCA cycle (Fig. 5.10), so that maximum conversion to glutamic acid can occur. To allow secretion of the glutamic acid, the cell envelope is made more permeable by certain procedures; this is necessary to prevent feedback-inhibition of glutamic acid synthesis.

Xanthan gum is an extracellular polysaccharide slime synthesized by strains of *Xanthomonas campestris*. It has many commercial uses—e.g. as a gelling agent, a gel stabilizer, a thickener and a crystallization inhibitor in various foods.

12.1.4 Vinegar

Vinegar is made by the 'acetification' of various ethanol-containing products—e.g. wine, cider, beer. Manufacture involves a carefully controlled process in which the ethanol is oxidized, aerobically, to acetic acid by species of *Acetobacter*. The vinegar may be aged, filtered, bottled and pasteurized.

'Vinegar' made chemically (e.g. by the carbonylation of methanol) is called 'non-brewed condiment'; it is cheaper than bacterially produced vinegar.

12.2   FOOD PRESERVATION

The various methods of food preservation aim to prevent or delay microbial and other forms of spoilage, and to guard against food poisoning; such methods therefore help the product to retain its nutritive value, extend its shelf-life and keep it safe for consumption. Physical methods of preservation (e.g. refrigeration) are generally preferred to chemical methods.

12.2.1   Physical methods of food preservation

12.2.1.1   *Pasteurization*

Pasteurization is a form of heat treatment used e.g. for milk, vinegar and certain foods; its object is to kill certain pathogens and spoilage organisms. Milk is held at a minimum temperature of 72 °C for at least 15 seconds (high-temperature, short-time [HTST] pasteurization). Pasteurization kills the causal agents of many milk-borne diseases (such as salmonellosis and tuberculosis) as well as much of the natural milk microflora; it also inactivates certain bacterial enzymes (e.g. lipases) which would otherwise cause spoilage. The causal agent of e.g. Q fever (*Coxiella burnetii*) is not necessarily killed by HTST pasteurization.

12.2.1.2   *Canning*

In the typical form of canning, suitably prepared foods are put into metal containers ('cans' or 'tins') which are then exhausted of air, sealed, and heated; the type of processing depends e.g. on the nature and pH of the food, but—in all cases—canned food must not contain *Clostridium botulinum* capable of growth and toxin production under the conditions of storage. Non-acid foods (e.g. potatoes, mushrooms) are heated sufficiently to destroy the endospores of *C. botulinum*. Additionally, canning inactivates organisms and enzymes capable of causing spoilage.

Once the cans have been sealed and properly processed, the food should remain safe for long periods of time (if stored correctly). The seal involves a double-seaming process (Fig. 12.1); very rarely, an imperfectly sealed can allows the ingress of microorganisms. Some organisms, e.g. certain endospore-formers such as *Bacillus stearothermophilus*, may survive the canning process because their heat-resistance is greater than that of *C. botulinum*; such organisms sometimes cause spoilage in canned foods.

12.2.1.3   *Refrigeration*

Temperatures such as 0–10 °C (used for short-term storage) can delay

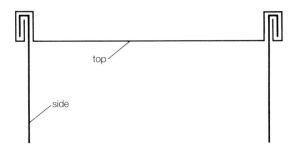

Fig. 12.1 Double-seaming, as used in the canning process (section 12.2.1.2). The edge of the can-end is wrapped around the body flange to form a seam of five thicknesses of metal. Not shown (but often present) are corrugations in the body and ends of the can; these help to relieve strain during the heating process.

spoilage by inhibiting the metabolism of contaminating organisms and/or the activity of their extracellular enzymes; even so, psychrotrophic organisms (section 3.1.4) may cause spoilage.

Freezing (used for long-term storage) may kill some contaminants, and it also reduces the amount of available water (section 3.1.3). At sub-zero temperatures—such as –5 to –10 °C—certain *fungi* may become important spoilage agents of e.g. meat.

12.2.1.4 *Dehydration*

Dehydration reduces the available water (section 3.1.3) to a point at which contaminants will not grow. The process may involve evaporation by heating—as e.g. in the manufacture of dried milk. Alternatively, the amount of available water may be reduced by adding sodium chloride (as in salted fish products) or syrups (as in preserved fruits).

12.2.1.5 *Ionizing radiation*

Ionizing radiation (e.g. high-energy electrons or *gamma*-radiation) is used in some countries for treating foods such as chicken and fish products, strawberries and spices.

12.2.2 Other methods of food preservation

12.2.2.1 *Acidification*

The traditional method of *pickling* preserves food by virtue of the lowered pH. This can be achieved either by adding acids (usually lactic acid, sometimes vinegar) or, in some cases, by fermenting the food; *sauerkraut* is cabbage which has been subjected to a natural lactic acid fermentation involving species of *Lactobacillus* and *Leuconostoc*.

12.2.2.2 Preservatives

Food preservatives are chemicals which can inhibit contaminants; some inhibit fungi as well as bacteria. They include e.g. benzoic acid, nitrites and sorbic acid. The use of preservatives is typically subject to governmental regulations.

13 Applied bacteriology II: miscellaneous aspects

13.1 FEEDING ANIMALS, PROTECTING PLANTS

As well as influencing soil fertility (Chapter 10), bacteria make specific contributions to agriculture by helping us to feed farm animals and assisting in the protection of certain crops.

13.1.1 Silage and single-cell protein

13.1.1.1 Silage

Silage-making, the traditional way of preserving grass (and certain other crops), enables animals to be fed during the winter months when vegetation is relatively scarce. Essentially, finely chopped grass is stored anaerobically and allowed to ferment. Bacteria—mainly *Lactobacillus* spp—are present on the vegetation and/or in the storage vessel (*silo*); they metabolize plant sugars (e.g. fructose, glucose, sucrose) mainly by a lactic acid fermentation (section 5.1.1.1). The production of lactic acid rapidly lowers the pH (to about 4.0), thereby inhibiting those organisms (particularly *Clostridium* spp) which would otherwise cause putrefaction; this helps to preserve the nutritive value of the crop. The acidity also helps to inhibit growth of *Listeria monocytogenes*, an organism which can grow in incompletely fermented silage (pH > 5.5); infection with *L. monocytogenes* may lead e.g. to septicaemia or abortion/stillbirth.

In 'big-bale' silage-making, the vegetation is fermented in large black plastic bags rather than in a silo.

13.1.1.2 Single-cell protein (SCP)

SCP refers to the cells of certain microorganisms (including bacteria, yeasts and microalgae) grown in large-scale cultures for use as a source of protein in the animal (and human) diet. In the early 1980s, when the price of protein was high, thousands of tons of bacterial protein were being manufactured each year for animal feed; the organism, *Methylophilus methylotrophus* (a methylotroph—section 6.4), was grown on a methanol substrate, and its yield was improved by incorporating an *E. coli* gene (encoding glutamate dehydrogenase) which improved the assimilation of ammonia.

13.1.2 Biological control

'Biological control' generally refers to the use, by man, of one species of organism to control the numbers or activities of another. Such exploitation is used on a commercial scale e.g. in agriculture and forestry. It commonly involves the use of certain microorganisms to kill or disable the insect pests of particular crop plants; such microorganisms are called 'microbial insecticides' or 'bioinsecticides'.

Strains of *Bacillus thuringiensis* which produce the δ-*endotoxin* are important bioinsecticides which are used world-wide against a range of insect pests on various crops. In some strains of *B. thuringiensis* this toxin is formed, in sporulating cells, as a *parasporal crystal* near the endospore; in other strains it seems to be formed only in non-sporing cells. Genes concerned with toxin synthesis appear to be encoded by a plasmid (section 7.1).

Preparations containing spores/crystals can be applied direct to crops. In the alkaline and proteolytic intestine of an insect larva, the crystal is converted to one or more toxic components which cause paralysis of the midgut and breakdown of the midgut epithelium; as a result the larva may die, or it may die subsequently of septicaemia.

13.2 BIOMINING

Chemolithotrophic bacteria, including species of *Thiobacillus* and *Sulfolobus*, are used to extract certain metals from low-grade ores; this process ('biomining' or 'bacterial leaching') is carried out mainly for the commercial recovery of copper from ores containing e.g. chalcopyrite ($CuFeS_2$) and iron pyrites (FeS_2). Essentially, a liquor containing sulphuric acid and chemo-lithotrophic bacteria is allowed to percolate through a mound of crushed ore, and is repeatedly re-cycled; ions leached from the ore (Fe^{2+}, S^{2-}) are oxidized by the bacteria, and the resulting Fe^{3+} and SO_4^{2-} solubilize further ferrous and cupric compounds from the ore. The process thus becomes self-sustaining. Various nutrients are provided by the liquor and/or by the ore itself. Copper in the leachate (up to 5 g/l) is periodically removed (e.g. by electrolysis). Bacterial metabolism maintains a temperature of about 50 °C, the convective upflow of air preserving the necessary aerobic conditions.

13.3 BIOLOGICAL WASHING POWDERS

'Biological' washing powders usually contain enzymes called *subtilisins*, produced by species of *Bacillus*. One of these, 'subtilisin Carlsberg' (obtained from *B. licheniformis*), can hydrolyse most types of peptide bond (in proteins) and even some ester bonds (in lipids). It is stable over a wide range of pH, and its stability does not depend on Ca^{2+}; this latter feature is important because washing powders often include agents which 'soften' water by chelating ions such as Ca^{2+}.

13.4   SEWAGE TREATMENT

Sewage includes domestic wastes (e.g. from drains and water-closets) and often varying amounts of agricultural and/or industrial effluent; it contains substances in suspension, in solution, and in colloidal form.

If discharged to rivers or lakes etc. sewage can be harmful in various ways. It can, for example, be a source of infection—promoting the spread of water-borne diseases such as cholera. Another problem is its content of dissolved organic matter; in metabolizing such nutrients, the large numbers of sewage bacteria can quickly use up the available oxygen in a locally polluted region—particularly in slow-moving or static waters. This can mean death for fish and other oxygen-dependent animals. Additionally, such anaerobiosis permits the growth of sulphate-reducing bacteria (section 5.1.1.2; Fig. 10.3) and other organisms whose metabolic products include sulphide and other malodorous substances. Two major aims of sewage treatment are therefore: (i) to eliminate (or reduce the numbers of) pathogens, and (ii) to diminish the oxygen-depleting ability of the sewage, i.e. to diminish its *biological oxygen demand* (BOD).

13.4.1   Aerobic sewage treatment

Aerobic treatment includes the familiar *trickle filter* (*biological filter*) in which sewage is sprayed, via holes in a horizontally rotating arm, onto a thick layer of crushed rock enclosed by a circular brick wall; percolating through the rock, the sewage makes close contact with surfaces that bear large numbers of ciliates (protozoa) and other organisms. With abundant oxygen available, dissolved organic substances are efficiently oxidized by sewage organisms and by those on the rock surfaces; the process is a controlled form of mineralization (section 10.3.4): after treatment, the effluent has a much lower BOD, i.e. when discharged to a river etc. it will take less oxygen from the water. As the sewage percolates, large numbers of bacteria are consumed by organisms on the rock surfaces.

13.4.2   Anaerobic sewage treatment

Sewage containing a relatively high content of solids—e.g. farm wastes, sludge from some aerobic treatment processes—can be treated by *anaerobic digestion*. In this process, complex organic matter is broken down to simple substances which include a high proportion of gaseous products; it involves a wide range of bacteria which, collectively, carry out a spectrum of metabolic activities. Essentially, sewage is digested in a tank at about 35 °C. Polymers, such as polysaccharides, are degraded by extracellular enzymes, and the resulting subunits (sugars etc.) are fermented (e.g. by species of *Bacteroides* and *Clostridium*) to products which include acetate, lactate, propionate, ethanol, CO_2 and hydrogen. Methanogens (section 5.1.2) produce methane from acetate and from the CO_2 and hydrogen; much of the bulk of sewage carbon is eliminated via CO_2 and methane.

Anaerobic digestion can yield a final product which is relatively odorless and rich in microbial biomass—a useful agricultural fertilizer. The gaseous product (*biogas, sewer gas*) may contain more than 50% of methane, and it can contribute most or all of the energy needs of the treatment process.

The technology for proper sewage treatment has been available (and used) for many years. Despite this, outfalls (long pipes) are still being built to convey raw (untreated) sewage into coastal waters; for those responsible, this may seem a 'cheap option', but for the *environment* it is certainly a most expensive option.

13.5 PUTTING PATHOGENS TO WORK

Pathogens such as *Clostridium botulinum* and *C. tetani* produce highly potent neurotoxins which cause severe or fatal disease. Botulinum toxin, for example, inhibits release of the neurotransmitter *acetylcholine*—with consequent reduction in muscle activity, or muscle paralysis. However, this precise effect of the toxin has been put to good use in the treatment of hyperactive muscle disorders such as strabismus ('squint'); treatment involves direct injection of the toxin into the muscle. The medical use of botulinum and other (microbial) neurotoxins has recently been reviewed [Schantz & Johnson (1992) MR 56 80–99].

14 Some practical bacteriology

14.1 SAFETY IN THE LABORATORY

A student new to bacteriology should be constantly aware that he or she is dealing with living organisms—which may include actual or potential pathogens. Good bacteriology is safe bacteriology, and it is wise to get to know the safety rules of the laboratory before carrying out any practical work; the following rules deserve special attention.

1. While working in the laboratory wear a clean laboratory coat to protect your clothing. Do not wear the coat outside the laboratory.
2. Put *nothing* into your mouth. It is potentially dangerous to eat, drink or smoke in the laboratory. For pipetting, use a rubber bulb (teat) or a mechanical device such as a 'pi-pump'; do not use your mouth for suction. If necessary, use self-adhesive labels.
3. Keep the bench—and the rest of the laboratory—clean and tidy.
4. Dispose of all contaminated wastes by placing them (not throwing them) into the proper container.
5. Leave contaminated pipettes, slides etc. in a suitable, active disinfectant for an appropriate time before washing/sterilizing them.
6. Avoid contaminating the environment with *aerosols* containing live bacteria/spores. An aerosol consists of minute (invisible) particles of liquid or solid dispersed in air; aerosols can form e.g. when bubbles burst, when one liquid is added to another, or when a drop of liquid falls onto a solid surface—things which can happen during many bacteriological procedures (see e.g. Fig. 16.2). Particles of less than a few micrometres in size can remain suspended in air for some time and may be inhaled by anyone in the vicinity; clearly, aerosols can be a potential source of infection. Bacteriological work is sometimes carried out in special cabinets (described later)—partly in order to avoid the risk of infection from aerosols.
7. Report all accidents and spillages, promptly, to the instructor or demonstrator.
8. Wash your hands thoroughly before leaving the laboratory.

14.2 BACTERIOLOGICAL MEDIA

A *medium* (plural: media) is any solid or liquid preparation made specifically for the growth, storage or transport of bacteria; when used for growth, the medium generally supplies all necessary nutrients. Before use, a medium must be *sterile*, i.e. it must contain no living organisms. (Methods for sterilizing media are given in Chapter 15.)

Before discussing the different media, it will be helpful to give again the meanings of a few words which are used very commonly in bacteriology; this is best done by giving the following outline of a simple laboratory procedure. To grow an organism such as *E. coli*, the bacteriologist takes an appropriate sterile medium and adds to it a small amount of material which consists of, or contains, living cells of that species; the 'small amount of material' is called an *inoculum*, and the process of adding the inoculum to the medium is *inoculation*. (The tools and procedures used for inoculation are described in sections 14.3 to 14.5.) The inoculated medium is then *incubated*, i.e. kept under appropriate conditions of temperature, humidity etc. for a suitable period of time. Incubation is usually carried out in a thermostatically controlled cabinet called an *incubator*. During incubation the bacteria grow and divide—giving rise to a *culture*; thus, a culture is a medium containing organisms which have grown (or are still growing) on or within that medium.

A liquid medium may be used in a test tube (which has a simple metal cap—see e.g. Fig. 16.3) or in a glass, screw-cap bottle; a *universal bottle* (Fig. 14.1) is a cylindrical bottle of about 25 ml capacity, while a *bijou* is smaller (about 5-7 ml).

Most solid media are jelly-like materials which consist of a solution of nutrients etc. 'solidified' by *agar* (a complex polysaccharide gelling agent obtained from certain seaweeds). A solid medium is commonly used in a plastic *Petri dish* (illustrated in Fig. 16.2)—usually the size which has a lid diameter of about 9 cm. The medium, in a molten (liquid) state is poured into the Petri dish and allowed to set; a Petri dish containing the solid medium is called a *plate*.

14.2.1   Types of medium

For many chemolithoautotrophic bacteria the medium can be a simple solution of inorganic salts (CO_2 being used for carbon).

Nutritionally undemanding heterotrophs (such as *E. coli*) need only the common organic substances found in *basal media* (Table 14.1). Many bacteria will not grow in basal media, but may do so after the addition of substances such as egg, serum or blood; media which have been supplemented in this way are called *enriched media*.

A *selective medium* is one which supports the growth of certain bacteria in preference to others. An example is MacConkey's broth (Table 14.1)—in which bile salts inhibit *non*-enteric bacteria but do not affect the growth of enteric species; this medium can be used e.g. to isolate enteric from non-

Table 14.1   Some common bacteriological media

Medium	*Composition of medium: typical formulation (% w/v in water)*
Basal medium	
Peptone water	Peptone (soluble products of protein hydrolysis) 1%; sodium chloride 0.5%
Nutrient broth[1]	Peptone 1%; sodium chloride 0.5%; beef extract 0.5–1%
Nutrient agar	Nutrient broth gelled with 1.5–2% agar
Differential medium	
MacConkey's agar	MacConkey's broth (see below) gelled with 1.5–2% agar
Enriched medium	
Blood agar	Nutrient agar (or similar medium) containing 5–10% defibrinated or citrated blood
Chocolate agar	Blood agar heated to 70–80 °C until the colour changes to chocolate brown
Serum agar	Nutrient agar (or similar medium) containing 5% (v/v) serum
Enrichment medium	
Selenite broth	Peptone 0.5%; mannitol 0.4%; disodium hydrogen phosphate 1%; sodium hydrogen selenite ($NaHSeO_3$) 0.4%
Selective medium	
Deoxycholate–citrate agar (DCA)	Meat extract and peptone 1%; lactose 1%; sodium citrate 1%; ferric ammonium citrate 0.1%; sodium deoxycholate 0.5%; neutral red (pH 8.0 yellow to pH 6.8 red) 0.002%; agar 1.5%
MacConkey's broth	Peptone 2%; lactose 1%; sodium chloride 0.5%; bile salts (e.g. sodium taurocholate) 0.5%; neutral red 0.003%
Transport medium	
Stuart's transport medium	Salts; agar 0.2–1.0% (semi-solid or 'sloppy' agar); sodium thioglycollate; methylene blue (as redox indicator)

[1]In bacteriology, 'broth' may refer to any of various liquid media, but, when used without qualification, it commonly refers to nutrient broth.

enteric bacteria when both types are present in an inoculum. (To some extent, *all* media are selective in that no medium can give equal support to the growth of every type of bacterium.)

An *enrichment medium* allows certain species to outgrow others by encouraging the growth of wanted organism(s) and/or by inhibiting the growth of unwanted species. Hence, if an inoculum contains only a few cells of the required species (among a large population of unwanted organisms), growth in a suitable enrichment medium can increase ('enrich') the proportion of required organisms. For example, selenite broth (Table 14.1) inhibits many types of enteric bacteria (including e.g. *E. coli*) but does not inhibit *Salmonella typhi*, the causal agent of typhoid. Suppose, for example, that we need to detect *S. typhi* in a specimen of faeces from a suspected case of typhoid. The specimen may contain only a few cells of *S. typhi*, so that it may be difficult or impossible to detect them among the vast numbers of non-pathogenic enteric bacteria. However, if an inoculum from the specimen is incubated in selenite broth, the proportion of cells of *S. typhi* increases to the point at which they can be detected more readily.

A *solid* medium is used, for example, to obtain the *colonies* (section 3.3.1) of a particular species. Many solid media are simply liquid media which have been solidified by a gelling agent such as gelatin or agar; agar is the most commonly used gelling agent because (i) it is not attacked by the vast majority of bacteria, and (ii) an agar gel does not melt at 37 °C—a temperature used for the incubation of many types of bacteria. (By contrast, gelatin can be liquefied by some bacteria, and it is molten at 37 °C.) One widely used agar-based medium is *nutrient agar* (Table 14.1), a general-purpose medium used for culturing (i.e. growing) many types of bacteria; it can also be enriched and/or made selective by the inclusion of appropriate substances.

Blood agar is an agar-based medium enriched with 5–10% blood; it is used e.g. for the culture (growth) of nutritionally 'fastidious' bacteria such as *Bordetella pertussis* (causal agent of whooping cough), and also to detect *haemolysis* (section 16.1.4.1). *Chocolate agar* is made by heating blood agar to 70–80 °C until it becomes chocolate brown in colour; it is more suitable than blood agar for growing certain pathogens (e.g. *Neisseria gonorrhoeae*).

MacConkey's agar is an example of a *differential medium*, i.e. one on which different species of bacteria may be distinguished from one another by differences in the characteristics of their colonies etc. On MacConkey's agar, lactose-utilizing enteric bacteria (such as *E. coli*) form *red* colonies because they produce acidic products (from the lactose) which affect the pH indicator in the medium; enteric species which do not use lactose (e.g. most strains of *Salmonella*) give rise to colourless colonies.

Some solid media contain neither agar nor gelatin. For example, *Dorset's egg* is made by heating (and, hence, coagulating) a mixture of homogenized hens' eggs and saline; it is used e.g. as a *maintenance medium*—i.e. a medium used first for the growth and then the 'storage' of a given organism. Dorset's egg has been widely used for the storage of *Mycobacterium tuberculosis*.

Many media contain substances (e.g. peptone, tap water) whose *exact* composition is usually unknown. Sometimes it is necessary to use a medium in which all the constituents, including those in trace amounts, are quantitatively known; such a *defined medium* is prepared from known amounts of pure substances—e.g. inorganic salts, glucose, amino acids etc. in distilled or de-ionized water. A defined medium would be used e.g. when determining the nutritional requirements of a given species of bacterium.

A *transport medium* is used for the transportation (or temporary storage) of material (e.g. a swab) which is to be subsequently examined for the presence of particular organism(s); the main function of the medium is to maintain the viability of those organism(s), if present. A transport medium need not support growth; in fact, growth may be disadvantageous since waste products formed may adversely affect the survival of the organisms. One such medium, Stuart's transport medium (Table 14.1), is suitable e.g. for a range of anaerobic bacteria and for 'delicate' organisms such as *Neisseria gonorrhoeae*.

14.2.2   The preparation of media

Most media can be obtained commercially in a dehydrated, powdered form. Such media are commonly dissolved in the appropriate volume of water, sterilized, and dispensed to suitable sterile containers; as an alternative, some media are dispensed to containers before sterilization.

For most agar-based media (e.g. nutrient agar) the powdered medium is mixed with water and steamed to dissolve the agar; the whole is then sterilized in an *autoclave* (section 15.1.1.3) and subsequently allowed to cool to about 45 °C, a temperature at which the agar remains molten. To prepare a *plate*, some 15–20 ml of the molten agar medium is poured into a sterile Petri dish which is left undisturbed until the agar sets. Blood agar plates are made by mixing molten nutrient agar (at about 45–50 °C) with 5–10%, by volume, of (e.g. citrated) blood before pouring the plates.

For some uses (e.g. streaking: section 14.5.2), the surface of a newly made plate must be 'dried'—i.e. *excess* surface moisture must be allowed to evaporate; this is often achieved by leaving the plate, with the lid partly off, in a 37 °C incubator for about 20 minutes. Spread plates (section 14.5.2) may also be dried.

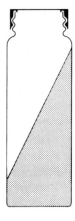

Fig. 14.1   A slope (also called a slant). The medium (stippled) has a large surface area available for inoculation; the thickest part of the medium is known as the *butt*. Slopes are commonly made of agar-based or gelatin-based media, and a slope is usually prepared in a universal bottle (as shown), in a bijou or in a test tube. Slopes are used e.g. for storing a purified strain of bacteria. A sterile slope is inoculated with an inoculum from a pure culture of the bacterial strain, and the slope is then incubated at a suitable temperature to allow growth; the slope can then be stored in a refrigerator at 4–6 °C until needed e.g. as a source of inoculum.

is allowed to set in a sterile bottle or test tube which has been placed at an angle to the horizontal.

Some types of medium cannot be sterilized by autoclaving because one or more of their constituents are destroyed at the temperatures reached in an autoclave. Such media include e.g. DCA (Table 14.l), which is steamed but not autoclaved, and those media which contain glucose or other heat-labile sugars; in preparing the latter type of medium the sugar solution is sterilized separately by filtration (section 15.1.3) before being added to the rest of the (autoclaved) medium.

14.3 ASEPTIC TECHNIQUE

Instruments and media etc. must, of course, be sterile before use; if we are not *sure* of their sterility we will simply not know what is happening in our practical work. Additionally, *during* bacteriological procedures, instruments and materials must be protected from contamination by organisms that are constantly present in the environment. *Aseptic technique* involves the pre-use sterilization of all instruments, vessels, media etc., and the avoidance of their subsequent contact with non-sterile objects—such as fingers, or the bench top etc.

Vessels containing sterile contents are kept closed except for the minimum time needed for introducing or removing material/inocula etc. Before opening a vessel (e.g. a sterile bottle, or one containing a pure culture), the rim of the screw-cap (or equivalent) is passed briefly through the bunsen flame to prevent any live contaminating organisms from falling into the vessel when the cap is removed; this procedure is called *flaming*, and it is used e.g. whenever an inoculum is withdrawn from a culture, or when a sterile medium is being inoculated. Flaming is repeated immediately before the vessel is closed. Flaming is generally not used when working with Petri dishes, and is never used when the contents of a vessel are likely to catch fire.

The risk of contamination in the laboratory may be further reduced by treating bench tops etc. with a suitable disinfectant, and by filtering the air to remove cells and spores of bacteria and fungi etc. Sometimes bacteriological work is done in a 'safety cabinet' (= 'sterile cabinet'). In a *class II* cabinet, sterile (filtered) air constantly flows down onto the work surface, and air passes to the exterior after further filtration (Fig. 14.2). Work is conducted via an open panel in the front of the cabinet. A *class III* cabinet is a gas-tight cabinet in which air is filtered before entry and before discharge to the environment; work is conducted via arm-length rubber gloves fitted into the front panel, and access to the interior of the cabinet is via a separate two-door sterilization/disinfection chamber. Class II cabinets are common in college laboratories; type III cabinets are used e.g. when working with highly pathogenic microorganisms.

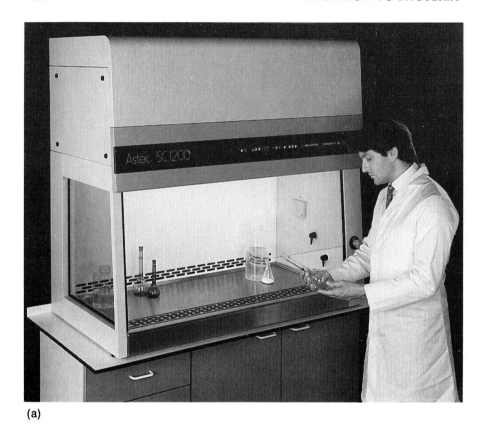

(a)

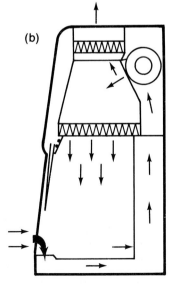

(b)

Fig. 14.2 (a) An Astec class II safety cabinet, and (b) the pattern of air-flow (arrows) during use; filtered air passes downwards onto the working surface, and air passes outwards through a filter at the top of the cabinet. (Courtesy of Astec Environmental Systems, Weston-super-Mare, Avon, UK.)

14.4   THE TOOLS OF THE BACTERIOLOGIST

In many cases bacteria can be handled with one of the simple instruments shown in Fig. 14.3. A loop or straight wire is sterilized immediately before use by flaming: the wire portion of the instrument is heated to red heat in a bunsen flame and is then allowed to cool.

If a sterile loop is dipped into a suspension of bacteria and withdrawn, the loop of wire retains a small circular film of liquid containing a number of bacterial cells—and this can be used as an inoculum; the size of this inoculum will depend on (i) the concentration of cells in the suspension, and (ii) the size of the wire loop (which often carries 0.01–0.005 ml of liquid)—clearly, two factors which can be controlled. Even smaller amounts of liquid can be manipulated with a straight wire since this picks up only the minute volume of suspension which adheres to the wire's surface.

The loop and straight wire can also be used for picking up small quantities of solid material—e.g. small amounts of growth from a bacterial colony—simply by bringing the wire loop, or the tip of the straight wire, into contact with the material; the *amount* of material which adheres to the wire will be unknown, but usually this is not important. Liquid or solid inocula carried by a loop or straight wire can be used to inoculate either a liquid or a solid medium (section 14.5).

Both the loop and straight wire must always be flamed immediately after use so that they do not contaminate the bench or environment. Spattering, with aerosol formation, may occur when flaming a loop or straight wire containing the residue of an inoculum; for this reason, flaming is often carried out with a special bunsen burner fitted with a tubular hood. Alternatively, flaming may be carried out in a safety cabinet.

Larger volumes of liquid may be handled by means of Pasteur pipettes or

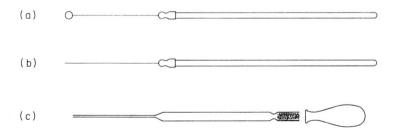

Fig. 14.3   Basic tools of the bacteriologist. (a) A loop: a piece of platinum, nickel-steel or nichrome wire, bent into a closed loop at the end and held in a metal handle of about 10–12 cm in length. (b) A straight wire: the metal handle carries a straight piece of wire of about 5–8 cm in length. (c) A Pasteur pipette: an open-ended glass tube, the narrow end of which has an internal diameter of about 1 mm; the wider end is plugged with cotton wool, before sterilization, and a rubber bulb (teat) is fitted immediately before use.

graduated pipettes; suction is obtained either from a rubber bulb (teat) or from a mechanical device—the mouth is never used. Pipettes used in bacteriology are usually plugged with cotton wool (Fig. 14.3), before being sterilized, in order to avoid contamination from the bulb or from the mechanical pipetting device during use. Pipettes are usually sterilized (in batches) inside metal canisters or in thick paper envelopes; when a pipette is removed from the container, only the plugged end should be held so as to avoid contaminating the rest of the pipette. Pasteur pipettes are commonly used once only and are then discarded into a jar of suitable disinfectant. Graduated pipettes which have been contaminated with bacteria are immersed in a disinfectant until they are safe to handle, when they can be washed up and re-used.

14.5   METHODS OF INOCULATION

14.5.1   Inoculating a liquid medium

To inoculate a liquid medium with a *liquid* inoculum, the loop (or straight wire) carrying the inoculum is simply dipped into the liquid medium, moved slightly, and then withdrawn. Inoculation can also be carried out with a Pasteur pipette. With a *solid* inoculum, the loop or straight wire may be rubbed lightly against the inside of the vessel containing the medium—to ensure that at least some of the inoculum is left behind when the instrument is withdrawn.

14.5.2   Inoculating a solid medium

Solid media may be inoculated in a variety of ways, particular methods being used for particular purposes.

Streaking (Fig. 14.4) is used when individual, well-separated colonies are required, and the (liquid or solid) inoculum is known to contain a large number of cells. In this method the inoculum is progressively 'thinned out' in such a way that individual, well-separated cells are left on at least some areas of the plate—usually in the third, fourth or fifth streakings (Fig. 14.4); on incubation, each well-separated cell gives rise to an individual colony.

In *stab inoculation*, a solid medium—e.g. the butt of a slope (Fig. 14.1)—is inoculated with a straight wire by plunging the wire vertically into the medium; the inoculum (at the tip of the wire) is thus distributed along the length of the stab. This procedure is used e.g. for inoculating deep, microaerobic or anaerobic parts of a medium.

A *spread plate* is made by spreading a small volume of liquid inoculum (e.g. 0.05–0.1 ml) over the surface of a solid medium by means of a sterile L-shaped glass rod (a 'spreader').

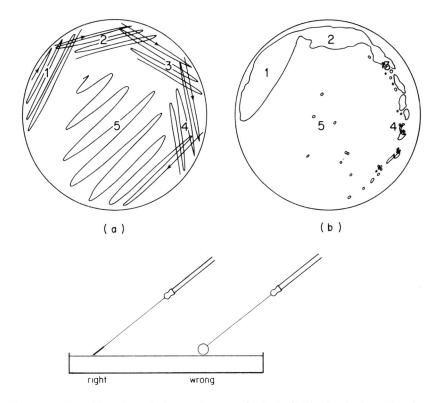

(a)   (b)

right   wrong

Fig. 14.4 Streaking: inoculating a plate to obtain individual colonies. (a) A loop carrying the inoculum is moved from side to side (i.e. streaked) across a peripheral region of the plate, following the path shown at 1. The loop is then flamed and allowed to cool; the *sterile* loop is now streaked across the medium as shown at 2. Streakings 3, 4 and 5 are similarly made, the loop being flamed and cooled between each streaking and after the last. (b) After incubation, those areas of the plate on which large numbers of cells had been deposited show areas of confluent growth, as at 1, 2 and 3; well-separated cells give rise to individual colonies, as at 5.

When streaking with a loop, the plane of the wire loop should *not* be vertical; starting from the wrong position (see lower diagram) the correct position is achieved by rotating the loop's handle through 90°.

A *flood plate* is made by flooding the surface of a solid medium with a liquid inoculum and withdrawing excess inoculum with a sterile Pasteur pipette.

If the inoculum in a flood plate, or a spread plate, contains enough cells, incubation will give rise to a *lawn plate*: a plate in which the surface of the medium is covered with a layer of confluent growth (section 3.3.1).

A plate is sometimes inoculated with a *swab*: typically, a compact piece of cotton wool attached securely to one end of a thin wooden stick or a piece of wire. A sterile swab is used e.g. for sampling organisms at a given site (such as the throat); after exposure, the swab is drawn lightly across the surface of a plate of suitable medium—taking care that all areas of the cotton wool make contact with the medium.

14.6   PREPARING A PURE CULTURE FROM A MIXTURE OF ORGANISMS

Some of the basic techniques of bacteriology can be illustrated by following through a common procedure such as the *isolation* of a particular strain or species from a mixture of organisms. The following describes the isolation of E. coli from a sample of sewage (which usually contains a range of enteric and non-enteric organisms).

A loopful of sewage is streaked onto a plate of MacConkey's agar (section 14.2.1), and the plate is then incubated for 18–24 hours at 37 °C. (Plates are incubated upside-down; if incubated the right way up, water vapour from the medium may condense on the inside of the lid and drop onto the surface of the medium—with possible disruption of the streaked inoculum.) During incubation, well-separated cells of *any* species capable of growing on the medium will each give rise to an individual colony. After 18–24 hours on MacConkey's agar, E. coli forms round, red colonies of about 2–3 mm in diameter—but not all colonies with this appearance will necessarily be those of E. coli. The next step is to choose several such colonies for further examination; since E. coli is very common in sewage, at least one of the selected colonies is likely to be that of E. coli.

Before identification can be attempted it is necessary to ensure that each of the selected colonies contains cells of only one species. There is always the possibility that a given colony—even a well-separated one—may contain the cells of two different species; this may occur if, during streaking, two different cells had been deposited (by chance) at the same point on the surface of the medium. To resolve this doubt, each colony is subcultured; *subculturing* is a process in which cells from an existing culture or colony are transferred to a fresh, sterile medium. To subculture a given colony, the surface of the colony is touched lightly with a sterile loop so that a minute quantity of growth adheres to the loop; the growth is then streaked (in this case) onto a plate of sterile MacConkey's agar. (Some bacteria form very small colonies, and in such cases it is often easier to subculture by touching the surface of the colony with the tip of a straight wire; the inoculum is then carried (on the straight wire) to a fresh medium where it is streaked with a sterile loop.) Each plate, inoculated from a single colony, is then incubated. If each red colony had been that of a single species, we should now have several pure cultures—at least one of which is likely to be that of E. coli. (To increase the chances that a culture is pure it may be subcultured again.) Each pure culture can now be examined by the identification procedures outlined in Chapter 16.

14.7   ANAEROBIC INCUBATION

Anaerobic bacteria are incubated under anaerobic conditions. This can be achieved by using an *anaerobic jar*—one form of which is the McIntosh and Fildes' jar: a strong, metal cylindrical chamber with a flat, circular, gas-tight lid. The jar is loaded with a stack of inoculated plates, and the lid is secured with a screw-clamp. The jar is then connected to a suction pump via one of two valves in the lid; after a few minutes the valve is closed. Hydrogen (in a rubber bladder) is then passed into the jar via the other valve; this valve is then closed. Evacuation and re-filling may be repeated several times. On the inside of the lid is a gauze envelope containing a catalyst (e.g. palladium-coated alumina pellets) which promotes chemical combination between hydrogen and the last traces of oxygen. The jar is then placed in an incubator for an appropriate period of time. (Within the jar, plates are stacked the right way up; if stacked upside-down, the agar may be sucked from the base of the Petri dish by the vacuum.)

Another (more modern) form of anaerobic jar is a stout cylindrical vessel of strong, transparent plastic with a flat, gas-tight lid. The jar is loaded with plates; water is then added to a small packet of chemicals which is dropped into the jar immediately before the lid is secured with a screw-clamp. The chemicals liberate hydrogen which, in the presence of a catalyst, combines with all the oxygen in the jar. Since, in this case, there is no vacuum, the plates can be inserted upside-down (i.e. lid-side down) so as to avoid the problem of condensation.

Most anaerobic jars contain a redox indicator which indicates the state of anaerobiosis in the jar. In metal jars the indicator is placed in a small glass side-arm, while in plastic jars an indicator-soaked pad is usually visible through the wall of the jar.

Some anaerobes can be grown (without an anaerobic jar) in media such as *Robertson's cooked meat medium* (minced beef heart, beef extract (1%), peptone (1%), sodium chloride (0.5%) and a reducing agent, e.g. L-cysteine or thioglycollate); the medium, sterilized by autoclaving, is stored in screw-cap universal bottles which are sometimes equilibrated under oxygen-free conditions before the cap is tightened.

14.8   COUNTING BACTERIA

The total number of (living and dead) cells in a sample is called the *total cell count*; the number of living cells is the *viable cell count*. Counts in liquid samples are usually given as the number of cells per millilitre (or per 100 ml).

(a)

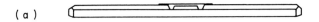

(b)

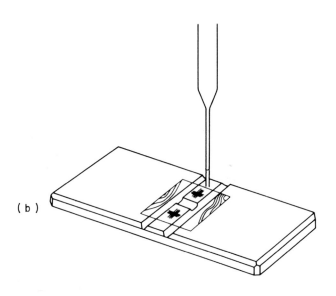

(c)

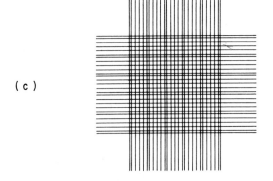

14.8.1 Total cell count

The total cell count in a liquid sample (e.g. a broth culture) can be estimated by direct counting in a counting chamber (Fig. 14.5).

Another method—the *direct epifluorescent filter technique* (DEFT)—is used e.g. for counting organisms in milk. Essentially, the milk is passed through a membrane filter, and the cells retained on the filter are stained with a fluorescent dye; ultraviolet radiation is then beamed onto the filter, and the (fluorescent) cells are seen (through a microscope) as bright particles against a dark background. (For DEFT, the milk is pre-treated to disrupt fat globules etc. which would otherwise block the filter.)

Fig. 14.5 A typical counting chamber (haemocytometer). The instrument, seen from one side at (a), consists of a rectangular glass block in which the central plateau lies precisely 0.1 mm below the level of the shoulders on either side. The central plateau is separated from each shoulder by a trough, and is itself divided into two parts by a shallow trough (seen at (b)). On the surface of each part of the central plateau is an etched grid (c) consisting of a square which is divided into 400 small squares, each 1/400 mm². A glass cover-slip is positioned as shown at (b) and is pressed firmly onto the shoulders of the chamber; to achieve proper contact it is necessary, while pressing, to move the cover-slip (slightly) against the surface of the shoulders. Proper (close) contact is indicated by the appearance of a pattern of coloured lines (Newton's rings), shown in black and white at (b).

Using the chamber. A small volume of a bacterial suspension is picked up in a Pasteur pipette by capillary attraction; the thread of liquid in the pipette should not be more than 10 mm. The pipette is then placed as shown in (b), i.e. with the opening of the pipette in contact with the central plateau, and the side of the pipette against the cover-slip. With the pipette in this position, liquid is automatically drawn by capillary attraction into the space bounded by the cover-slip and part of the central plateau; *the liquid should not overflow into the trough*. (It is sometimes necessary to tap the end of the pipette, *lightly*, against the central plateau to encourage the liquid to enter the chamber.) A second sample can be examined, if required, in the other half of the counting chamber. The chamber is left for 30 minutes to allow the cells to settle, and counting is then carried out under a high power of the microscope—which is focused on the grid of the chamber. Since the volume between grid and cover-slip is accurately known, the count of cells per unit volume can be calculated.

A worked example. Each small square in the grid is 1/400 mm². As the distance between grid and cover-slip is 1/10 mm, the volume of liquid over each small square is $1/4000^3$ mm—i.e. 1/4,000,000 ml.

Suppose, for example, that on scanning all 400 small squares, 500 cells were counted; this would give an average of $500 \div 400$ (= 1.25) cells per small square, i.e. 1.25 cells per 1/4,000,000 ml. The sample therefore contains $1.25 \times 4,000,000$ cells/ml, i.e. 5×10^6 cells/ml.

If the sample had been diluted before examination (because it was too concentrated), the count obtained must be multiplied by the dilution factor; for example, if diluted 1-in-10, the count should be multiplied by 10.

N.B. The chamber described above is the *Thoma chamber*; in a *Helber chamber* the distance between central plateau and cover-slip is 0.02 mm.

The total cell count can also be estimated by comparing the *turbidity* of the sample with that of each of a set of tubes (*Brown's tubes*) containing suspensions of barium sulphate in increasing concentration; the tubes range from transparent (tube 1), through translucent, to turbid and opaque (tube 10). For a given species of bacterium the turbidity of a particular tube corresponds to the turbidity of a suspension of cells of known concentration. The sample is examined in a tube of size and thickness equivalent to those containing the standard suspensions; the turbidity of the sample is matched, visually, with that of a particular tube, and the concentration of the sample is then read from a table supplied with the tubes.

14.8.2   Viable cell count

Most methods of estimating the viable cell count involve the inoculation of a solid medium with the sample (or diluted sample). After incubation, the number of cells in the inoculum can be estimated from the number of colonies which develop on or within the medium. It is always assumed that each colony has arisen from a single cell; the number of cells which actually give rise to colonies depends at least partly on the type of medium used and on the conditions of incubation.

In the *spread plate* (or *surface plate*) method, an inoculum of about 0.05–0.1 ml is spread over the surface of a sterile plate, as described earlier; the plate is 'dried' (section 14.2.2), incubated, and the viable cell count is estimated from (i) the number of colonies, (ii) the volume of inoculum used, and (iii) the degree (if any) to which the sample had been diluted. If a sample is suspected of containing many cells—e.g. 10^6 cells/ml—it can be diluted in 10-fold steps, and an inoculum from each dilution spread onto a separate plate; at least one dilution will give a countable number of colonies.

In the *pour plate* method, the (liquid) inoculum is mixed with a molten agar-based medium (at about 45 °C) which is then poured into a Petri dish and allowed to set; on incubation, colonies develop within (as well as on) the medium, and the viable count is calculated as in the spread plate method.

Yet another method for viable count is *Miles and Misra's method* (Fig. 14.6).

A sample likely to contain small numbers of bacteria (e.g. water from a *clean* river) may be passed through a sterile membrane filter of pore size about 0.2 μm—which retains cells on the upper surface; a volume of, say, 100 ml or more may be filtered. The membrane is then placed (cell-side uppermost) onto an absorbent pad saturated with a suitable medium; on incubation, nutrients diffuse through the membrane, and colonies develop from those cells capable of growth under such conditions. The viable cell count can then be estimated from (i) the number of colonies on the membrane, and (ii) the volume of sample filtered.

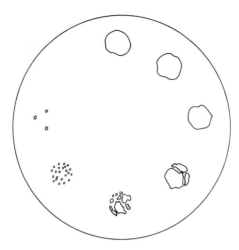

Fig. 14.6 Miles and Misra's method (the 'drop method') for estimating viable cell counts. The sample is first diluted, usually in 10-fold steps—i.e. 1/10, 1/100 . . . etc. One drop (of known volume) from each dilution is then placed at a separate, recorded position on the surface of a 'dried' plate of suitable medium. The drops are allowed to dry, and the plate is incubated until visible growth develops. Drops which contained large numbers of viable cells give rise to circular areas of confluent growth. Any drop which contained less than about 15 viable cells will give rise to a small, countable number of colonies; by assuming that each colony arose from a single cell, the viable count can be estimated from (i) the number of colonies, (ii) the volume of the drop, and (iii) the dilution factor.

The same Pasteur pipette can be used throughout *if* the drops are taken first from the highest dilution and then from the next highest dilution, and so on.

A Pasteur pipette can be *calibrated*, i.e. the volume of drops delivered by the pipette can be measured. This is done by drawing into the pipette a measured volume (e.g. 1 ml) of water, and counting the number of drops formed when this volume is discharged; if 1 ml is discharged in, say, 30 drops, then each drop is about 1/30 ml.

14.8.3 Counting cells in (or on) solids

Sometimes we need to count the bacteria in a sample of food or fabric etc. In order to do this, the particles of food etc. must be broken up and/or the bacteria must be separated from the sample; ideally, clumps of bacteria should also be broken up. The sample and a diluent (such as Ringer's solution) can be sealed into a sterile plastic bag (a *stomacher*) which is then subjected to mechanical agitation; in this way, at least some of the bacteria can be brought into suspension in the diluent.

14.9 STAINING

Staining is often used to detect, categorize or identify bacteria, or to observe specific bacterial components; in most cases the cells are killed and 'fixed' before being stained.

Dyes etc. are usually applied to a thin film of cells on a glass microscope slide. Cells from a pure culture may be examined as follows. A loopful of water is placed on a clean slide, and (using the loop)a speck of growth from a colony is mixed ('emulsified') with the water to form a suspension of cells. Using the loop, the suspension is spread over an area of one or two square centimetres and allowed to dry—forming a *smear*. The smear is then fixed by passing it quickly through a bunsen flame twice; it is then ready for staining and subsequent examination under the microscope.

A smear may also be made directly from centrifuged urine, or pus from an abscess.

14.9.1   The Gram stain

The background to this stain is given in section 2.2.9. One of many versions of the procedure is as follows.

A heat-fixed smear (see above) is stained for 1 minute with *crystal violet*; it is then rinsed briefly under running water, treated for 1 minute with *Lugol's iodine* (a solution of iodine and potassium iodide in water), and briefly rinsed again. Decolorization is then attempted by treating the stained smear with a solvent such as ethanol (95%), acetone or iodine–acetone. This is the critical stage: with the slide tilted, the solvent is allowed to run over the smear only for as long as dye no longer runs *freely* from it (about 1–3 seconds); the smear is then *immediately* rinsed in running water. At this stage, any Gram-negative cells will be colourless; Gram-positive cells will be violet. The smear is now counterstained for 30 seconds with dilute *carbolfuchsin* to stain (red) any Gram-negative cells present. After a brief rinse, the smear is blotted dry and examined under the oil-immersion objective of the microscope (magnification about 1000×).

Certain species of bacteria do not give a clear or constant reaction to the Gram stain—sometimes reacting positively, sometimes negatively; these bacteria are said to be *Gram-variable*. To avoid the problems of Gram-variability in taxonomy (classification), many bacteria are described as *Gram type positive* or *Gram type negative* according to whether their cell walls are of the Gram-positive or Gram-negative type, respectively (section 2.2.9).

14.9.2   The Ziehl–Neelsen stain (acid-fast stain)

'Acid-fast' bacteria differ from all other bacteria in that once they are stained with hot, concentrated carbolfuchsin they cannot be decolorized by mineral acids or by mixtures of acid and ethanol; such bacteria include e.g. *Mycobacterium tuberculosis*. A heat-fixed smear is flooded with a concentrated solution of carbolfuchsin, and the slide is heated until the solution steams; it should not boil. The slide is kept hot for about 5 minutes, left to cool, and then rinsed in running water. Decolorization is attempted by passing the slide through several changes of acid–alcohol (e.g. 3% v/v concentrated hydrochloric acid in 90% ethanol). After washing in water, the smear is counterstained with a contrasting stain (such as malachite green), washed again, and dried. Acid-fast cells stain red, others green.

14.9.3 Capsule stain

Bacterial capsules may be demonstrated e.g. by *negative staining* (Plate 2: centre, right). The cells are emulsified with a loopful of e.g. *nigrosin* solution on a clean slide and overlaid with a cover-slip; under the oil-immersion or 'high dry' objective lens of the microscope the capsule appears as a clear, bright zone between a cell and its dark background.

14.9.4 Endospore stain

See section 16.1.1.4.

15 Man against bacteria

Bacteria can be a nuisance, or even dangerous, in many everyday situations, and we therefore need methods to eliminate them or to inhibit their activities. Sometimes it is necessary to destroy, completely, all forms of life on a given object—as, for example, when surgical instruments are prepared for use. At other times it may be sufficient merely to eliminate only the potentially harmful organisms. There is also the special problem of inactivating pathogenic bacteria on or within living tissues.

15.1 STERILIZATION

Any procedure guaranteed to kill *all* living organisms—including endospores (section 4.3.1) and viruses—is called a *sterilization* process. (Since sterilization kills *all* organisms, expressions such as 'partial sterilization' are meaningless.) Ideally, sterilization methods should be efficient, quick, simple and cheap, and they should be applicable to a wide range of materials; sterilization is usually carried out by physical methods—commonly by the use of heat.

15.1.1 Sterilization by heat

The cells of different species of bacteria vary in their susceptibility to heat, and endospores are much more resistant than vegetative cells; vegetative cells generally die rapidly in boiling water, while endospores may survive for long periods of time.

The sterilizing power of heat depends not only on temperature but also on factors such as time, the presence of moisture, and the number and condition of the microorganisms present.

15.1.1.1 Fire

Fire is used e.g. for the rapid sterilization of surfaces and loops etc. (sections 14.3, 14.4), while disposable items such as used surgical dressings and one-use syringes may be sterilized—and destroyed—by incineration. However, less destructive methods are used for most other purposes.

15.1.1.2 The hot-air oven

This apparatus is used e.g. for the sterilization of heat-resistant items such as clean glassware. In use, a temperature of 160–170 °C is maintained for 60–90 minutes; this denatures proteins, desiccates cytoplasm and oxidizes various components in any organisms present. Within the oven, air should be circulated by a fan to ensure that all parts are kept at the required temperature; items should be well spaced in order not to impede the flow of air.

15.1.1.3   Sterilization by steam: the autoclave

Steam can sterilize at lower temperatures (for shorter times) than those used in a hot-air oven. At normal atmospheric pressure, steam has a temperature of only 100 °C—a temperature at which some endospores can survive for long periods—but, when *under pressure*, steam can reach higher temperatures suitable for sterilization; in fact, there is a definite relationship between the pressure and temperature of *pure* steam, i.e. steam containing no air: the higher the pressure the higher the temperature. When sterilizing with steam, items to be sterilized are placed inside a strong, metal, gas-tight chamber (an *autoclave*). Steam is produced within the chamber (Fig. 15.1) or, in larger autoclaves, is piped in from a boiler; air passes out through a valve until the chamber is filled with pure steam—at which time the valve is closed. The pressure and temperature of the steam rise as heating is continued (Fig. 15.1) or as more steam is piped in. At a pre-determined pressure, an (adjustable) valve opens—thus determining the pressure/temperature within the autoclave; steam which escapes (via the valve) is replaced by steam generated in the chamber, or piped in, so that pressure in the autoclave remains constant.

The time allowed for sterilization must be sufficient for all parts of the *load* (i.e. items/materials being sterilized) to reach the sterilizing temperature and to stay at that temperature until any organisms present have been killed.

Temperature/time combinations commonly used in autoclaves are (i) 115 °C (that is, a pressure of 72 kPa [10 lbs/inch^2] higher than atmospheric pressure) for 35 minutes; (ii) 121 °C (108 kPa [15 lbs/inch^2]) for 15–20 minutes; (iii) 134 °C (217 kPa [30 lbs/inch^2]) for 4 minutes.

For effective sterilization the steam must be saturated, i.e. it must hold as much water in vapour form as is possible for the given temperature and pressure; no air should be present because air upsets the pressure-temperature relationship: an air–steam mixture at a given pressure has a lower temperature than that of pure steam at the same pressure. Hence, *all* air must be purged from the chamber—and from all items within the chamber—before the valve is closed.

Small portable laboratory autoclaves generally resemble the domestic pressure cooker both in principle and mode of use (Fig. 15.1). In larger autoclaves, such as those in hospitals, steam is usually piped to the autoclave chamber from a boiler, and factors such as timing, pressure and steam

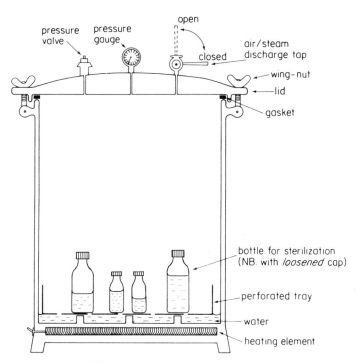

Fig. 15.1 A typical, small laboratory autoclave. Water is placed in the bottom of the chamber. Objects for sterilization are placed on the perforated tray which holds them above the water. The lid, with the air/steam discharge tap *open*, is clamped securely in position; the rubber gasket ensures a gas-tight seal. The heating element is switched on and the water boils. Steam fills the chamber, eventually displacing all the air (which leaves via the discharge tap). Pure steam begins to issue vigorously from the discharge tap, which is then closed. As heating continues, water continues to vaporize so that the pressure (and hence temperature) within the chamber increases. Once the desired pressure/temperature has been reached (see text), a pre-set pressure valve opens; steam escapes, thus maintaining the pressure at a steady level. When the appropriate time has elapsed (see text), the heating element is switched off and the autoclave is allowed to cool until the pressure inside the chamber (indicated by the gauge) does not exceed atmospheric pressure; the lid can then be opened safely and the sterile contents removed.

quality are often controlled automatically. In some models steam is admitted at the top of the chamber so that air is displaced downwards; this is more effective than upward displacement (used in small autoclaves) since steam is lighter than air under these conditions.

In another type of autoclave, air is removed from the chamber by a vacuum pump before steam is admitted; this allows rapid and thorough penetration by the steam of porous materials such as dressings or bed linen—materials which tend to trap air.

Some materials cannot be sterilized by autoclaving; these include water-repellent substances (e.g. petroleum jelly) and substances which are volatile or are heat-labile (i.e. destroyed by heat). Some of these materials (such as petroleum jelly) can be sterilized in a hot-air oven.

Certain materials, which would be damaged by autoclaving, may be sterilized by steam at reduced pressure (at e.g. 80°C) together with formaldehyde; this method kills endospores within about 2 hours, and is used for sterilizing heat-sensitive surgical instruments, plastic tubing, woollen blankets etc.

15.1.2 Sterilization by ionizing radiation

Ionizing radiation—e.g. *beta*-rays (electrons), *gamma*-rays, X-rays—sterilizes by supplying energy for a variety of lethal chemical reactions in the contaminating organisms. *Gamma*-radiation (typically using a cobalt-60 source) is widely used e.g. for the sterilization of pre-packed biological equipment such as plastic Petri dishes and syringes.

15.1.3 Sterilization by filtration

A liquid known to be free of the smallest microorganisms (e.g. viruses and subviral agents) may be sterilized by passing it through a *membrane filter* (of suitable pore size) which can retain all other microorganisms; the liquid may be drawn through the filter by reduced pressure in the (sterile) receiving vessel, or forced through e.g. by a syringe plunger. The filter itself consists of a thin sheet of cellulose acetate, polycarbonate or similar material; the pore size may be e.g. 0.2 μm.

Filtration is used e.g. for the sterilization of serum (for laboratory use), solutions of heat-labile antibiotics and media containing heat-labile sugars.

15.1.4 Sterilization by chemical agents

Chemicals used for sterilization (*sterilants*) are necessarily highly reactive and damaging to living tissues; they therefore require careful handling, and tend to be used only in larger institutions with suitable equipment and personnel.

Ethylene oxide (C_2H_4O)—a water-soluble cyclic ether—is a gas at temperatures above 10.8°C and forms explosive mixtures with air; it is therefore used diluted with another gas such as carbon dioxide or nitrogen. For sterilization, the gas mixture is used in a special chamber, and the temperature, humidity, time, and concentration of ethylene oxide must be carefully controlled. Ethylene oxide is an alkylating agent which reacts with various groups in proteins and nucleic acids; it is used e.g. for sterilizing clean medical equipment, bed linen, and heat-labile materials such as certain plastics.

Other sterilants include glutaraldehyde and β-propiolactone.

15.2   DISINFECTION

Disinfection is a procedure which destroys, inactivates or removes *potentially harmful* microbes—without necessarily affecting the other organisms present; it generally has little or no effect on bacterial endospores. 'Disinfection' often refers specifically to the use of certain chemicals (*disinfectants*) for the treatment of non-living objects or surfaces, but the term is sometimes also used to refer to antisepsis (section 15.3). Although chemical disinfection is widely used, physical methods are more suitable for certain purposes.

15.2.1   Disinfection by chemicals

Ideally, disinfectants for general use should be able to kill a wide range of common or potential pathogens. However, any given disinfectant is usually more effective against some organisms than it is against others, and the activity of a disinfectant may be greatly affected by factors such as dilution, temperature, pH, or the presence of organic matter or detergent; to be effective at all, a disinfectant needs appropriate conditions, at a suitable concentration, for an adequate period of time. Some disinfectants (e.g. hypochlorites) tend to be unstable, and some (e.g. 'pine disinfectants') need solubilization in order to be effective. At low concentrations some disinfectants not only cease to be effective, they can actually be metabolized by certain bacteria—e.g. species of *Pseudomonas* can grow in dilute solutions of 'carbolic acid' (phenol).

Disinfectants which *kill* bacteria are said to be *bactericidal*. Others merely halt the growth of bacteria, and if such a disinfectant is inactivated—e.g. by dilution, or by chemical neutralization—the bacteria may be able to resume growth; these disinfectants are said to be *bacteriostatic*. A bactericidal disinfectant may become bacteriostatic when diluted.

Of the many disinfectants in use, only a few of the common ones are mentioned here.

Phenol and its derivatives (e.g. 'phenolics' such as *cresols* and *xylenols*) can be bactericidal at appropriate concentrations; they appear to act mainly by affecting the permeability of the cytoplasmic membrane. *Lysol* is a mixture of methylphenols solubilized by soap; at 0.5% it kills many non-sporing pathogens in 15 minutes, but endospores may survive in 2% Lysol for days.

Chlorine is widely used for the disinfection of water supplies and for the sanitation of water in swimming pools. It acts (directly, and via hypochlorous acid) as an effective disinfectant, though its activity is decreased by the presence of organic matter and by other substances with which it can react.

Quaternary ammonium compounds (QACs) are cationic detergents used e.g. for the disinfection of equipment in the food and dairy industries. They are bacteriostatic at low concentrations, bactericidal at higher concentrations, and are typically more active against Gram-positive than Gram-negative bacteria. QACs appear to disrupt the cytoplasmic membrane; their activity is inhibited e.g. by soaps, some cations (e.g. Ca^{2+}, Mg^{2+}), low pH and organic matter.

Hypochlorites are highly effective against a wide range of bacteria (including endospores), the undissociated form of HOCl being strongly bactericidal. Sodium hydroxide is commonly used as a stabilizer in commercial hypochlorite disinfectants.

15.2.2 Disinfection by physical agents

Ultraviolet radiation can damage DNA and can be lethal to bacteria under appropriate conditions. It has poor powers of penetration (being readily absorbed by solids), but ultraviolet lamps (wavelength about 254 nm) are used e.g. for the disinfection of air and exposed surfaces in enclosed areas.

The disinfection of milk by *pasteurization* is discussed in section 12.2.1.1.

15.3 ANTISEPSIS

Antisepsis is the disinfection of *living* tissues; it may be used prophylactically (i.e. to prevent infection) or therapeutically (i.e. to treat infection).

The comments on disinfectants (section 15.2.1) generally apply also to *antiseptics*, i.e. chemicals used for antisepsis.

Dettol is a general-purpose phenolic antiseptic when used in dilute form, but a domestic disinfectant in more concentrated form; it is based on chloroxylenols.

Hexachlorophene has been used in antiseptic soaps; it is a *bisphenol* (i.e. the molecule contains two phenolic groups) which is much more effective against Gram-positive than Gram-negative bacteria.

A 70:30 *ethanol*:water mixture is used as a general skin antiseptic.

Soaps generally have little or no antibacterial activity unless they contain antiseptics; however, soap can help to remove bacteria from the skin—along with dirt and grease.

QACs (section 15.2.1) include cetyltrimethylammonium bromide (*Cetrimide*, *Cetavlon* etc.) which is used in antiseptic creams.

Iodine (in alcoholic or aqueous solution) is a potent bactericidal and sporicidal antiseptic.

15.4   ANTIBIOTICS

Originally, 'antibiotic' meant any microbial product which, even at very low concentrations, inhibits or kills certain microorganisms; the term is now generally used in a wider sense to include, in addition, any semi-synthetic or wholly synthetic substance with these properties.

No antibiotic is effective against all bacteria. Some are active against a narrow range of species, while others are active against a *broad spectrum* of organisms—including both Gram-positive and Gram-negative bacteria. In some cases a natural antibiotic (one produced by a microbe) can be chemically modified in the laboratory to form a 'semi-synthetic' antibiotic which may have a significantly different spectrum of activity.

Like disinfectants, antibiotics can be either bactericidal or bacteriostatic, and one which is bactericidal at one concentration may be bacteriostatic at a lower concentration.

An antibiotic characteristically acts at a precise site in the cell; depending on antibiotic, the site of action may be in the cell wall, the cytoplasmic membrane or the protein-synthesizing machinery, or it may be an enzyme involved in nucleic acid synthesis. Since a bacterium differs in many ways from a eukaryotic cell (Table 1.1), the toxic effect of an antibiotic on a bacterium is unlikely to be exerted e.g. on human or animal cells; because of this *selective toxicity*, some antibiotics are useful for treating certain diseases: a bacterial pathogen can be attacked without harming the host. Clearly, any antibiotic used in this way must retain activity within the body for long enough to be effective against the pathogen.

Of the many known antibiotics, relatively few are suitable for treating disease; some of these are mentioned briefly below.

15.4.1   β-Lactam antibiotics

These antibiotics include the penicillins, cephalosporins, carbapenems, clavams and monobactams; in each case the molecule includes a four-membered nitrogen-containing ring, the *β-lactam ring*. In some of the antibiotics the β-lactam ring is susceptible to cleavage by enzymes (*β-lactamases*) produced by certain bacteria; such cleavage destroys the antibiotic. β-Lactam antibiotics act by disrupting synthesis of the cell envelope in growing cells: they inactivate penicillin-binding proteins and inhibit the synthesis of peptidoglycan (sections 2.2.8 and 6.3.3.1); *only* growing cells are killed.

15.4.1.1   *Penicillins*

The original penicillins (e.g. *benzylpenicillin*) have low activity against Gram-negative bacteria owing to poor penetration of the outer membrane (section 2.2.9.2); they also have little or no effect against those Gram-positive bacteria which form β-lactamases. Some semi-synthetic penicillins (such as *cloxacillin, methicillin, nafcillin*) are resistant to a range of different β-lactamases (including those formed by staphylococci), but they are still poorly effective against Gram-negative bacteria. *Ampicillin* and its derivatives (e.g. *amoxycillin*) combine resistance to some β-lactamases with increased activity against Gram-negative bacteria.

15.4.2   Aminoglycoside antibiotics

These broad-spectrum, typically bactericidal antibiotics include *amikacin, gentamicin, kanamycin, neomycin* and *streptomycin*; they are active against both Gram-positive and Gram-negative bacteria. Aminoglycoside antibiotics act by binding to one or both subunits of the ribosome and inhibiting e.g. translocation in protein synthesis (Fig. 7.9).

15.4.3   Tetracyclines

These broad-spectrum antibiotics inhibit protein synthesis by binding to ribosomes and inhibiting the binding of aminoacyl-tRNAs to the 'A' site (Fig. 7.9); they are used for treating human and animal diseases caused e.g. by *Brucella, Chlamydia, Mycoplasma* and *Rickettsia*—and, interestingly, for the treatment of certain plant diseases such as coconut lethal yellowing (caused by a *Mycoplasma*-like organism).

15.4.4   Chloramphenicol

A broad-spectrum, bacteriostatic agent which binds to the 50S ribosomal subunit and inhibits transpeptidation in protein synthesis (Fig. 7.9).

15.4.5   Polymyxins

Polymyxins are peptides which are active against many Gram-negative bacteria; most Gram-positive bacteria are resistant. In Gram-negative bacteria; it inhibits e.g. the supercoiling activities of *gyrase*, an enzyme involved in DNA replication (section 7.3).

15.4.6   Nalidixic acid

This wholly synthetic antibiotic acts against a range of Gram-negative bacteria; it inhibits e.g. the supercoiling activities of *gyrase*, an enzyme involved in DNA replication (section 7.3).

15.4.7 Novobiocin

An antibiotic, produced e.g. by *Streptomyces niveus*, which is more active against Gram-positive than Gram-negative bacteria; it inhibits DNA replication by inhibiting the supercoiling activities of gyrase.

15.4.8 Rifamycins

These antibiotics are generally active against Gram-positive bacteria (including mycobacteria and staphylococci) and against some Gram-negative bacteria; they inhibit RNA polymerase, thus inhibiting synthesis of the 'leading' strand in DNA replication (section 7.3) and of mRNA in protein synthesis (section 7.6).

15.4.9 Sulphonamides

These synthetic compounds are bacteriostatic for a wide range of Gram-positive and Gram-negative bacteria; they interfere with the synthesis of *folic acid*: a coenzyme essential in a number of vital metabolic reactions. A sulphonamide molecule is similar in shape to a normal component of folic acid: *p*-aminobenzoic acid (PABA); during folic acid synthesis, sulphonamide may be incorporated instead of PABA—resulting in the formation of an inactive analogue of folic acid.

15.4.10 Synergism and antagonism between antibiotics

If two or more different antibiotics act simultaneously on an organism, such that each contributes to the death/inactivation of the organism, the antibiotics are said to be acting *synergistically*. Some antibiotics act *antagonistically*; for example, those which inhibit growth (e.g. chloramphenicol) will antagonize the β-lactam antibiotics—which act only when cells are growing.

15.4.11 Bacterial resistance to antibiotics

Why are some bacteria not affected by some antibiotics? In some cases a bacterium is resistant because it lacks the target structure of a given antibiotic; for example, species of *Mycoplasma* (which lack cell walls) will not be affected by penicillins—whose ultimate target (peptidoglycan) is a cell-wall component. Some bacteria may not carry out the particular process affected by an antibiotic: sulphonamides, for example, will not affect organisms which normally obtain their folic acid, ready-made, from the environment. Resistance can also be due to the ability of a cell to exclude an antibiotic from the target site; in many Gram-negative bacteria the outer membrane is impermeable to certain antibiotics, and in both Gram-positive and Gram-negative species the cytoplasmic membrane may act as a barrier.

Some bacteria can produce enzyme(s) which inactivate particular antibiotics: certain strains of *Staphylococcus*, for example, produce penicillinases—enzymes which can inactivate certain penicillins; penicillinases are examples of β-lactamases (section 15.4.1).

Resistance to antibiotics can also be *acquired*. This can happen e.g. through mutation (section 8.1) or by the acquisition of an R plasmid (section 7.1). A mutation may alter the target site so that it is no longer affected by the antibiotic; for example, a mutation conferring resistance to streptomycin may alter the ribosome so that streptomycin no longer binds to it, or, if binding does occur, ribosomal function is not affected. A single mutation usually confers resistance to only one antibiotic, or to closely related antibiotics which have the same target site.

An R plasmid, or a transposon (section 8.3), may encode resistance to one or to many related or unrelated antibiotics. It may, for example, encode enzymes (e.g. β-lactamases) which inactivate particular antibiotics; in some cases, the genes encoding these enzymes are inducible (section 7.8). A different form of acquired resistance is that mediated by the *TET protein* (encoded by transposon Tn*10*); this protein, which is induced in the presence of tetracyclines, appears to act like a shuttle, transporting tetracycline molecules outwards across the cytoplasmic membrane.

15.4.11.1 *Antibiotic-sensitivity tests*

Tests are often carried out to determine the susceptibility of a pathogen to a range of antibiotics; the results of such tests may enable the clinician to select optimally active antibiotic(s) for chemotherapy (section 11.9), and the pattern of sensitivity to antibiotics may also be useful in identifying the pathogen.

One common form of test is the *disc diffusion test*. A plate of suitable agar medium is inoculated from a pure culture of the pathogen; the entire surface of the medium is inoculated (often with a swab) so that a lawn plate (section 14.5.2) will develop on incubation. Before incubation, several small absorbent paper discs, each impregnated with a different antibiotic, are placed at different locations on the inoculated medium; on subsequent incubation, antibiotics diffuse from the discs, and a zone of growth-inhibition develops around each disc containing an antibiotic to which the organism is sensitive. Methods must be standardized.

16 The identification and classification of bacteria

16.1 IDENTIFICATION

How is an unknown species of bacterium identified? Usually, the first step is to determine certain of its characteristics; these are then compared with the characteristics of each of a number of known, named species until a 'match' is found. The principle is simple enough, but, in practice, which characteristics are determined—and must the unknown species be compared with each of the thousands of known species of bacteria? Fortunately, the source of an unknown species often gives clues which, together with a few simple observations and tests, may indicate the possible identity of the organism, or, at least, serve to narrow the search to one of the major groups of bacteria. For example, if the bacterium comes from faeces, and is found to be a Gram-negative, motile, facultatively fermentative bacillus, the bacteriologist would immediately think of the family Enterobacteriaceae—since this family contains many bacteria of that type, some of which are common in faeces. Identification to species level may then be possible by comparing the characteristics of the unknown species with those of genera and species of the Enterobacteriaceae. Clearly, the practice of identification is made easier if the bacteriologist (i) has a knowledge of the types of organism likely to be present in a given environment, and (ii) is familiar with the main distinguishing features of the families, genera and species of common bacteria—information of the type given in the Appendix.

In order to carry out tests and observations on an unknown species, it is commonly necessary to start with a pure culture of the organism (section 14.6); if a test be carried out on an impure culture (a mixture of organisms) the reaction of one organism may conflict with that of another so that, generally, it will be impossible to obtain a meaningful result. However, even with a pure culture it is occasionally found that the characteristics of the unknown strain of bacterium do not match, exactly, those of any species in a manual of identification. This can occur, for example, if the organism is a mutant strain (section 8.1) in which the mutation has altered one or more of the organism's 'typical' characteristics; similarly, a strain containing a plasmid (section 7.1) may display characteristics which are not typical of the species to which it belongs. Nevertheless, the organism is not always to blame: 'inexplicable' results can sometimes be due to slight variations in the methods and/or material used in the tests themselves.

In identifying an unknown bacterium (obtained in pure culture) the following characteristics are often the first to be determined since they have the greatest *differential* value, each helping to exclude one or more of the major groups of bacteria: (i) reaction to certain stains, particularly the Gram stain (section 14.9.1); (ii) morphology (coccus, bacillus etc.); (iii) motility; (iv) the ability to form endospores; (v) the ability to grow under aerobic and/or anaerobic conditions; (vi) the ability to produce the enzyme *catalase*. Fortunately, these characteristics are among the easiest to determine.

16.1.1 Preliminary observations and tests

16.1.1.1 Reaction to stains

A smear (section 14.9) from a pure culture is generally Gram-stained (section 14.9.1). A capsule stain (section 14.9.3) is used when capsulation is an important differential feature.

16.1.1.2 Morphology

Morphology is generally determined by examining a stained smear under the microscope. The smear may be stained either by Gram's method or by a simpler procedure—for example, by treating a heat-fixed smear for one minute with methylene blue or carbolfuchsin. Usually the stained smear is examined under the oil-immersion objective of the microscope (total magnification about 1000×), although the cells of some species (e.g. *Bacillus megaterium*) are clearly visible under the 'high-dry' lens (total magnification about 400×).

16.1.1.3 Motility

Motility (section 2.2.15.1) can often be determined by examining a 'hanging drop' preparation (Fig. 16.1) under the microscope. Even with unstained cells and ordinary (bright-field) microscopy, it is often possible to see whether or not the cells are motile; however, cells can be seen more clearly with dark-field or phase-contrast microscopy. Motility should be distinguished from *Brownian motion*: small, random movements exhibited by any small particle of about 1 μm or less when freely suspended in a liquid medium; these movements (seen e.g. in particles of colloidal clay) are due to bombardment of the particles by molecules of the liquid.

Motility can sometimes be inferred from the way an organism grows on solid media: motile species may tend to spread outwards from the inoculated area as organisms swim in the thin layer of surface moisture.

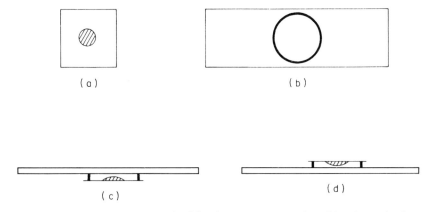

Fig. 16.1   The hanging-drop method for determining motility. (a) A drop of culture containing live, unstained bacteria is placed on a clean cover-slip. (b) A ring of plasticine is pressed onto a microscope slide. (c) The slide is inverted and pressed onto the cover-slip. (d) The whole is inverted for examination under the microscope.

16.1.1.4   Endospore formation

Relatively few bacteria can form endospores (section 4.3.1). If an endospore-former is grown for several days or a week on a solid medium, endospores can usually be detected by treating a heat-fixed smear of the growth with a 'spore stain'. This resembles the Ziehl–Neelsen stain (section 14.9.2) but uses e.g. ethanol for decolorization; vegetative cells are decolorized but endospores retain the (red) dye. Vegetative cells may be counterstained. (The shape of an endospore and its position within the cell are features which are sometimes used in identification.)

Endospores may be detected indirectly by heating a culture to 80 °C for 10 minutes—a procedure which endospores usually survive but which kills most types of vegetative cell. Any growth following subculture to a fresh medium suggests the presence of endospores.

16.1.1.5   Aerobic/anaerobic growth

Whether an organism is an aerobe, anaerobe or facultative anaerobe (section 3.1.6) is easily determined by attempting culture aerobically and (section 14.7) anaerobically.

16.1.1.6   Catalase production

Catalase is an iron-containing enzyme which catalyses the decomposition of hydrogen peroxide (H_2O_2) to water and oxygen; it is formed by most aerobic bacteria, and it de-toxifies the hydrogen peroxide produced during aerobic

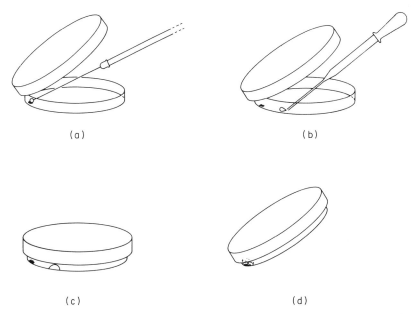

Fig. 16.2   The catalase test: a method which avoids contamination of the environment by aerosols. (a) A small quantity of bacterial growth is placed in a clean, empty, non-vented Petri dish. (b) Two drops of hydrogen peroxide are placed in the Petri dish a short distance from the growth. (c) The Petri dish is closed. (d) The closed Petri dish is tilted so that the hydrogen peroxide runs onto the bacterial growth. A positive reaction is indicated by the appearance of bubbles. The lid can be taped to the base before disposal in the proper container.

metabolism. The catalase test is used to detect the presence of catalase in a given strain of bacterium. Essentially, hydrogen peroxide is added to bacterial cells (or vice versa), the presence of catalase being indicated by bubbles of gas (oxygen). In the traditional form of the test, a speck of bacterial growth is transferred, with a loop, to a drop of hydrogen peroxide on a slide; however, in a positive test the bursting of bubbles will give rise to an aerosol (section 14.1). The author's method (Fig. 16.2) avoids this problem.

Some bacteria (e.g. certain strains of *Lactobacillus* and *Streptococcus faecalis*) produce *pseudocatalase*, a non-iron-containing enzyme which behaves like catalase.

If growth used for the catalase test is obtained from a colony on blood agar (section 14.2.1), care should be taken to exclude erythrocytes (red blood cells) from the sample since they contain catalase—and may therefore give a false-positive result.

16.1.2   Secondary observations and metabolic ('biochemical') tests

Once the search for identity has been narrowed to one or a few families, the bacteriologist uses some simple 'biochemical' tests; these tests distinguish between bacteria of different genera and species by detecting differences in their metabolism. For example, a test may distinguish between species which can and cannot ferment a particular carbohydrate, or which produce different products when metabolizing a particular substrate. The following are a few (of many) tests which are frequently carried out in bacteriological laboratories.

16.1.2.1   The oxidase test

This test detects a particular type of respiratory chain (section 5.1.1.2): one containing a terminal cytochrome c and its associated *oxidase*. Bacteria which contain such a chain can oxidize chemicals such as *Kovács' oxidase reagent* (1% tetramethyl-*p*-phenylenediamine dihydrochloride); electrons are transferred from this reagent to cytochrome c and thence, via the oxidase, to oxygen. When oxidized in this way, the reagent develops an intense violet colour. In the test, a small area of filter paper is moistened with a few drops of Kovács' oxidase reagent, and a small amount of bacterial growth is smeared onto the moist filter paper with a glass spatula or a platinum loop (but *not* a nichrome loop); oxidase-positive species give a violet coloration immediately or within 10 seconds. Oxidase-positive bacteria include e.g. species of *Neisseria*, *Pseudomonas* and *Vibrio*; a negative reaction is given e.g. by members of the family Enterobacteriaceae. (The used test paper should be disposed of safely.)

16.1.2.2   The coagulase test (for staphylococci)

Some ('coagulase-positive') strains of *Staphylococcus* produce one or (usually) both of two different proteins called *coagulases*; this test is used to detect coagulase production.

Free coagulase (= true coagulase, or staphylocoagulase) is released into the medium and is detected (in a *tube test*) by its ability to coagulate (i.e. clot) plasma containing an anticoagulant such as citrate, oxalate or heparin; the anticoagulant is necessary since, without it, the plasma would clot spontaneously. (Note that some bacteria can metabolize citrate, so that a false-positive reaction, i.e. spontaneous clotting, may occur if citrated plasma is used for these organisms.) In one form of tube test, 0.5 ml of an 18–24-hour broth culture of the strain under test is added to 1 ml of plasma in a test-tube; the tube is kept at 37 °C and examined for the presence of a clot after 1, 2, 3 and 4 hours, and at 24 hours. Free coagulase triggers conversion of the plasma protein *fibrinogen* to *fibrin*—which forms the clot.

Those coagulase-positive strains which also produce a *fibrinolysin* (an enzyme which lyses fibrin) may not form a clot, or may lyse a clot soon after its formation—hence the need for frequent examination of the tube.

Bound coagulase (= clumping factor) is a protein component of the cell surface; it binds to fibrinogen, resulting in the clumping of cells (*para-coagulation*). (Contrast this with the clotting of plasma.) Bound coagulase is detected by a *slide test*: a loopful of citrated or oxalated plasma is stirred into a drop of thick bacterial suspension on a microscope slide; in a positive test, cells clump within 5 seconds.

Known coagulase-positive and coagulase-negative strains should be used as controls in each form of the test.

Some other bacteria (e.g. strains of *Yersinia pestis*) also form coagulases.

16.1.2.3   The oxidation–fermentation (O–F) test

This test (= Hugh & Leifson's test) determines whether an organism uses oxidative (respiratory) or fermentative metabolism for the utilization of a given carbohydrate (usually glucose). Two test-tubes are filled to a depth of about 8 cm with a peptone–agar medium containing the given carbohydrate and a pH indicator, bromthymol blue, which makes the medium green (pH 7.1). One of the tubes is steamed (to remove dissolved oxygen) and is quickly cooled just before use. Each tube is then stab-inoculated (section 14.5.2) with the test organism to a depth of about 5 cm; in the 'steamed' tube the medium is immediately covered with a layer of sterile liquid paraffin about 1 cm deep (to exclude oxygen). Both tubes are then incubated and later (1–14 days) examined for evidence of carbohydrate utilization, namely, acid-production—indicated by yellowing of the pH indicator.

Respiratory organisms (such as *Pseudomonas* species) cause yellowing only in the uncovered ('aerobic') medium. *E. coli* causes yellowing in both media; glucose is fermented in the covered medium, and is attacked first by respiration and then by fermentation in the uncovered medium.

16.1.2.4   Acid/gas from carbohydrates ('sugars')

In some genera the species can be distinguished from one another by differences in the types of carbohydrate ('sugar') which they can metabolize. The range of sugars utilized by any particular organism can be determined simply by growing the organism in a series of media, each containing a different sugar and a system for detecting sugar utilization. The medium may be based on peptone-water or nutrient broth: it contains, in addition to the sugar, a pH indicator to detect acidification due to metabolism of the sugar. Such 'peptone-water sugars' or 'broth sugars' are used e.g. in tests on bacteria of the family Enterobacteriaceae. If a particular sugar is metabolized, acid products will be formed, and the acidity is detected by the pH indicator.

Certain bacteria cannot be tested in media containing peptone-water or broth—for example, species of *Bacillus* may form excess alkaline products (from the peptone or broth) so that any acid formed from sugar metabolism would not be detected; for some species the tests may be carried out in media containing inorganic salts and a given sugar. For other types of bacteria the medium must be enriched with e.g. serum, otherwise growth will not occur.

Test media are generally used in test-tubes, or in bijoux (section 14.2), and the tube or bijou may contain an inverted Durham tube (Fig. 16.3) to collect gas that may be formed during the metabolism of the sugar. Gas may be formed e.g. in the mixed acid and butanediol fermentations (Figs 5.5 and 5.6), formic acid being split into carbon dioxide and hydrogen by the *formate hydrogen lyase* enzyme system.

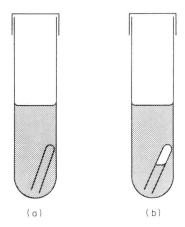

(a)   (b)

Fig. 16.3   The detection of gas produced during growth in a liquid medium. (a) An uninoculated tube of liquid medium containing an inverted Durham tube. (b) A gas-producing organism has been grown in the medium; some of the gas has collected in the Durham tube.

16.1.2.5 IMViC tests

The IMViC tests are a group of tests used particularly for identifying bacteria of the family Enterobacteriaceae. 'IMViC' derives from: *i*ndole test, *m*ethyl red test, *V*oges–Proskauer test and *c*itrate test.

The indole test. This test determines the ability of an organism to produce indole from the amino acid tryptophan. The organism is grown in peptone-water or tryptone-water for 48 hours; to the culture is then added Kovács' indole reagent (0.5 ml per 5.0 ml culture) and the (closed) container is gently shaken. In a positive test, indole (present in the culture) dissolves in the reagent—which then becomes pink, or red, and forms a layer at the surface of the medium.

The methyl red test (MR test). The MR test determines the ability of an organism, growing in a phosphate-buffered glucose–peptone medium, to produce sufficient acid (from the metabolism of glucose) to reduce the pH of the medium from 7.5 to about 4.4 or below. The medium is inoculated and is then incubated for at least 48 hours at 37 °C, following which the pH of the culture is tested by adding a few drops of 0.04% methyl red (yellow at pH 6.2, red at pH 4.4); with an MR-positive organism the culture becomes red.

The Voges–Proskauer test (VP test). This test detects the ability of an organism to form acetoin (acetylmethylcarbinol)—see butanediol fermentation, Fig. 5.6. A phosphate-buffered glucose–peptone medium is inoculated with the test strain and incubated at 37 °C for 2 days, or at 30 °C for at least 5 days. In one form of the test (*Barritt's method*), 0.6 ml of an ethanolic solution of 5% α-naphthol, and 0.2 ml of 40% potassium hydroxide solution, are added sequentially to 1 ml of culture; the (stoppered) tube or bottle is then shaken vigorously, placed in a sloping position (for maximum exposure of the culture to air), and examined after 30 and 60 minutes. Acetoin (if present) is apparently oxidized to diacetyl ($CH_3.CO.CO.CH_3$) which, under test conditions, gives a red coloration (a positive VP test).

The citrate test. This test determines the ability of an organism to use citrate as the sole source of carbon. Media used for the test—e.g. Koser's citrate medium (a liquid), and Simmons' citrate agar—include citric acid or citrate, ammonium dihydrogen phosphate (as a source of nitrogen and phosphorous, sodium chloride and magnesium sulphate. A saline suspension of the test organism is made from growth on a solid medium; using a *straight wire* (section 14.4), Koser's medium is inoculated from the suspension and is then incubated and examined for signs of growth (turbidity) after one or two days. Organisms which grow in the medium are designated 'citrate-positive'. A straight wire is used so that little or no nutrient is carried over in the inoculum; any nutrient carried over from the original medium may permit a small amount of growth in the citrate medium, thus giving a false-positive result. In an alternative method of inoculation, a straight wire is used to transfer a small quantity of growth from the *top* of a colony direct to the test medium.

16.1.2.6   *Hydrogen sulphide production*

Many species of bacteria produce hydrogen sulphide, e.g. by the reduction of sulphate (section 5.1.1.2) or from the metabolism of sulphur-containing amino acids. A sensitive test for sulphide is likely to be positive even for those species which form very small amounts of sulphide. A test of *low* sensitivity can distinguish between those species which form negligible or small amounts of sulphide and those which form large amounts. In one form of test, the organism is stab-inoculated into a tube of solid, gelatin-based medium containing peptone and a low concentration of ferrous chloride; organisms which form a lot of sulphide form visible amounts of black ferrous sulphide. In a more sensitive test, a strip of lead acetate paper is placed above the medium on/in which the test organism is growing; hydrogen sulphide production is indicated by the formation of lead sulphide, which causes blackening of the strip.

16.1.2.7   The urease test

Ureases are enzymes which hydrolyse urea, $(NH_2)_2.CO$, to carbon dioxide and ammonia. Urease production in enterobacteria can be detected by culture on e.g. *Christensen's urea agar*: a phosphate-buffered medium containing glucose, peptone, urea, and the pH indicator phenol red (yellow at pH 6.8, red at pH 8.4); when grown on this medium, 'urease-positive' strains liberate ammonia which raises the pH and causes the pH indicator to turn red.

16.1.2.8   Decarboxylase tests

These tests determine the ability of an organism to form *decarboxylases*, enzymes which decarboxylate the amino acids arginine, lysine and ornithine to agmatine, cadaverine and putrescine, respectively. Three tubes of *Møller's decarboxylase broth*, each including glucose, peptone, one of the amino acids, and the pH indicators bromcresol purple and cresol red, are inoculated with the test organism; each broth is covered with a layer of sterile paraffin (to exclude air), incubated at 37 °C, and examined daily for 4 days. Initially the medium becomes acidic (yellow) due to glucose metabolism; if a decarboxylase is *not* formed the medium remains yellow. Decarboxylation of the amino acid produces an alkaline product which subsequently raises the pH, causing the medium to become purple. A *control* medium resembles the test medium but lacks an amino acid; it should become, and remain, yellow.

16.1.2.9   The phenylalanine deaminase test (PPA test)

This test determines the ability of an organism to deaminate phenylalanine to phenylpyruvic acid (PPA). The organism is grown overnight on phenylalanine agar (containing yeast extract, Na_2HPO_4, sodium chloride and DL-phenylalanine); 0.2 ml of a 10% solution of ferric chloride is then added to the growth. PPA, if present, gives a green coloration with the ferric chloride. PPA-positive bacteria include e.g. *Proteus vulgaris*.

16.1.2.10   The ONPG test

The utilization of lactose often involves two enzymes: (i) a galactoside 'permease' (which facilitates the uptake of lactose), and (ii) β-D-galactosidase (which splits lactose into glucose and galactose); species such as *E. coli* can usually synthesize both of these enzymes. Certain bacteria which do not utilize lactose, or which metabolize it very slowly, may nevertheless form the enzyme β-D-galactosidase; the inability of such organisms to carry out normal lactose metabolism may be due, for example, to an inability to synthesize galactoside permease. To detect the presence of β-D-galactosidase in such organisms, use is made of a substance, *o*-nitrophenyl-β-D-galactopyranoside (ONPG), which can enter the cell without a specific permease; once inside the cell, ONPG is cleaved by the galactosidase to galactose and the yellow-coloured *o*-nitrophenol. In the ONPG test, the organism is grown for 18–24 hours in broth containing ONPG; a positive test (β-D-galactosidase present) is indicated by the appearance of the (yellow) *o*-nitrophenol in the medium.

16.1.2.11   The phosphatase test

Phosphatases, enzymes which hydrolyse organic phosphates, are produced by a number of bacteria and can be detected by the phosphatase test. The organism is grown for 18–24 hours on a solid medium which includes sodium phenolphthalein diphosphate; this substance is hydrolysed by phosphatases with the liberation of phenolphthalein—a pH indicator which is colourless at pH 8.3 and red at pH 10.0. To detect phenolphthalein (a positive test), the culture is exposed to gaseous ammonia, which causes phosphatase-containing colonies to turn red.

16.1.2.12   The nitrate reduction test

This test determines the ability of an organism to reduce nitrate (see e.g. anaerobic respiration, section 5.1.1.2). The following test can be used e.g. for enterobacteria and pseudomonads. The organism is grown for one or more days in nitrate broth (e.g. peptone water containing 0.1–0.2% w/v potassium nitrate), and the medium is then examined for evidence of nitrate reduction. To test for *nitrite*, 0.5 ml of 'nitrite reagent A' and 0.5 ml of 'nitrite reagent B' are added to the culture; these reagents combine with any nitrite present to form a soluble red azo dye. The *absence* of red coloration could mean that either (i) nitrate had not been reduced, or (ii) nitrite was formed but had been subsequently reduced e.g. to nitrogen or ammonia. To distinguish between these two possibilities, the medium is tested for the presence of nitrate by adding a trace of zinc dust—which reduces nitrate to nitrite; if nitrate is present (i.e. it has not been reduced by the test organism), the addition of zinc will bring about a red coloration since the newly-formed nitrite will react with the reagents present in the medium.

16.1.2.13 Reactions in litmus milk

Many species of bacteria give characteristic reactions when they grow in *litmus milk* (skim-milk containing the pH indicator litmus). A given strain of bacterium may produce one or more of the following effects: (i) no visible change; (ii) acid production from the milk sugar (lactose) indicated by the litmus; (iii) alkali production, usually due to hydrolysis of the milk protein (casein); (iv) reduction (decolorization) of the litmus; (v) the production of an acid clot, which is soluble in alkali; (vi) the formation of a clot at or near pH 7 due to the action of rennin-like enzymes produced by the bacteria; (vii) the production of acid *and* gas which may give rise to a *stormy clot*: a clot which has been disturbed and is permeated by bubbles of gas.

16.1.3 Micromethods

In clinical bacteriology, 'micromethods' are miniaturized test procedures which are used to carry out, simultaneously, a range of routine biochemical identification tests. These procedures involve the use of commercial 'kits' which save time, space and materials; a few are mentioned below.

The *API* system consists of a plastic strip holding a number of microtubes, each containing a different dehydrated medium; each microtube is inoculated from a suspension of the test organism, mineral oil is added to certain microtubes (to exclude air), and the strip is incubated. Later, reagents are added, where appropriate, to detect particular metabolic products.

Enterotube II consists of a tube divided into a sequence of 12 compartments, each containing a different agar-based medium; the media are inoculated by passing an inoculum-bearing straight wire axially through the tube.

The *PathoTec* system consists of various test strips, each impregnated with the dehydrated medium appropriate to a given test; each strip is incubated in a suspension of the test organism (or inoculated directly from a colony) and the test is read after a specified time.

16.1.4 Miscellaneous observations and tests

16.1.4.1 Haemolysis

When certain bacteria grow on blood agar (section 14.2.1), each colony is surrounded by a 'halo' of differentiated medium in which the erythrocytes have been lysed or in which the blood has been discoloured; this phenomenon is called *haemolysis*, and it is caused by the action of substances (*haemolysins*) released by the bacteria. Some species produce glass-clear, colourless haemolysis which contrasts sharply with the opaque red medium; this is formed e.g. by *Streptococcus pyogenes* and by some strains of *Staphylococcus aureus*. (When formed by *Streptococcus*, glass-clear haemolysis is often called β-haemolysis, but when formed by *S. aureus* it may be referred to as α-haemolysis; it is probably best to refer to this type of haemolysis simply as 'clear haemolysis'.)

The so-called 'viridans streptococci', and some other bacteria, form zones of greenish-brown discoloration (*greening*) around their colonies.

For any given haemolytic bacterium, haemolysis—or a particular form of it—may develop only if the organism has been grown on media containing the blood of specific type(s) of animal (e.g. horse, rabbit, man etc.).

16.1.4.2   Serological tests

These tests can distinguish between closely related bacteria which differ in their cell-surface *antigens* (section 11.4.2). For example, using serological tests, thousands of different strains of *Salmonella* can be distinguished primarily by slight chemical differences in their O antigens (lipopolysaccharide–protein antigens) and H antigens (flagellar antigens); differences in the O antigens, for example, occur in the O-specific chains (section 2.2.9.2). Strains which are distinguished mainly on the basis of their antigens are called *serotypes* (see e.g. *Salmonella* in the Appendix).

In practice, we can detect (and identify) a given serotype by using specific *antibodies* which are known to combine only with the antigens of known serotype(s). Antibodies are obtained by injecting an experimental animal with antigens from a known serotype; after an interval of time, the animal's serum will contain antibodies to those antigens. Such a serum is called a specific *antiserum*. On mixing this antiserum with cells of the given serotype, combination will occur between cell-surface antigens and their corresponding antibodies in the antiserum; when this happens, the bacterium–antibody complexes form a visible whitish suspended mass, or a sediment, in the test-tube (an *agglutination reaction*). If this same antiserum can agglutinate an unidentified strain, it can be concluded that the unknown strain has antigen(s) in common with the original serotype. Hence, an unknown serotype can be identified by testing it with each of a range of antisera, each antiserum containing antibodies of particular, known serotype(s).

16.1.4.3   Bacteriophage typing ('phage typing')

This procedure distinguishes between different strains of closely related bacteria by exploiting differences in their susceptibility to a range of bacteriophages (Chapter 9). A *flood plate* (section 14.5.2) is prepared from a culture of a given strain; the plate is 'dried' (section 14.2.2), and a grid is drawn on the base of the Petri dish. Next, the agar over each square of the grid is inoculated with one drop of a suspension of phage—each square being inoculated with a different phage; the drops are allowed to dry and the plate is incubated. Usually, one, two or more of the phages will be found to be lytic for a given strain; lysis (susceptibility to a given phage) is indicated by the formation of a plaque (section 9.1.3) at the point of inoculation of that phage on the flood plate. In this way, strains can be defined (and identified) by the range of phages to which they are susceptible.

16.2   CLASSIFICATION

Ideally, biological (*taxonomic*) classification should reflect natural, evolutionary relationships between organisms. Bacteria, however, have been traditionally classified on the basis of criteria which do not necessarily have evolutionary significance (section 1.3); even so, such classification has provided useful 'keys' for identification purposes, and has permitted a logical system of naming.

Since the 'uniqueness' of any bacterium depends ultimately on its genome (Chapters 7 and 8), taxonomically useful differences between bacteria have been sought, and found, in their nucleic acids; some of these differences have been used as a basis for a form of classification which is believed, by many, to reflect evolutionary relationships among the bacteria.

16.2.1   GC ratios (GC%)

In chromosomal DNA, the *GC ratio* (GC%) is:

(guanine + cytosine)/(guanine + cytosine + adenine + thymine)%

GC% values vary according to genus and species, thus providing a useful taxonomic criterion; however, similar values among strains do not necessarily indicate a close taxonomic relationship, though widely differing values would indicate the absence of such a relationship.

The GC% of a DNA sample can be estimated by determining its 'melting point', i.e. the temperature at which the strands of dsDNA separate; the melting point varies—linearly—with GC%, so that the GC% of a given sample of DNA can be found by interpolation from a graph in which melting point has been plotted against GC%. Alternatively, GC% can be estimated by specialized centrifugation techniques.

16.2.2   DNA fingerprinting in classification

In this procedure chromosomal DNA is isolated from each of the strains to be compared. The DNA from each strain is then exposed to specific restriction endonuclease(s) (section 7.4; Table 8.1), and the resulting fragments of DNA are separated by gel electrophoresis; the fragments are denatured *in situ* (i.e. in the gel), and are then transferred from the gel to a nitrocellulose (or other) medium by the technique of *Southern blotting*. Once blotted, the fragments can be stained. The strains can then be compared by comparing their individual patterns of stained fragments; clearly, identical strains will give identical patterns, but different patterns will be formed from different types of chromosmal DNA.

Owing to the denaturing process, the DNA which is transferred to the blotting medium is single-stranded. As an alternative to staining, the pattern of single-stranded fragments can be investigated with one or more labelled *probes* (section 8.5.2) for sequences of particular interest.

16.2.3   rRNA oligonucleotide cataloguing

As mentioned earlier (section 2.2.3), sequences of nucleotides in rRNA are believed to have been generally stable over evolutionary periods of time; consequently, any significant change in these sequences may indicate evolutionary divergence. Such evidence may be useful e.g. for the construction of a classification scheme based on real (evolutionary) relationships.

In this procedure, rRNA (particularly 16S rRNA) from a given organism is cleaved, enzymatically, into small pieces (oligonucleotides) in each of which the nucleotide sequence is determined; this gives a 'catalogue' which is characteristic of the organism. Different organisms can be classified, on the basis of their catalogues, into groups which may have phylogenetic (evolutionary) significance.

On the basis of 16S rRNA cataloguing, two major groups of bacteria have been distinguished: the Archaebacteria and the Eubacteria (see Appendix). rRNA analysis has also been used for intra-genus classification—e.g. for grouping many of the species of *Pseudomonas*.

Organisms which have diverged very recently (on an evolutionary time-scale) may not be recognized as separate species on the basis of their 16S rRNA sequences. For example, certain species of *Bacillus*, which can be distinguished from one another by DNA–DNA homology studies (i.e. comparison of their chromosomal DNA), have been shown to have virtually identical 16S rRNAs [Fox, Wisotzkey & Jurtshuk (1992) IJSB 42 166–170]; this underlines the stability of 16S rRNA, and suggests that its sequence may change only very late in the evolution of a species. .

Appendix   Minidescriptions of some genera, families, orders and other categories of bacteria

The following minidescriptions give essential features of many of the bacteria mentioned in the text; they are intended for the rapid orientation of the reader. Further details of these and of many other bacteria (and other microorganisms) are given in *Dictionary of Microbiology & Molecular Biology* [Singleton & Sainsbury (2nd edition, 1987) published by John Wiley & Sons, Chichester]; entries in the dictionary also include terms, tests, techniques, biochemical pathways, and topics in genetics, molecular biology, medicine and immunology.

Unless specifically indicated, details given for any particular category cover all the organisms in that category; hence, for example, 'motile or non-motile' means that some member(s) of that category are motile and other(s) are non-motile.

The GC% (section 16.2.1) gives the range for the genus or other category.

The 'type species' is the species which is regarded as the 'permanent representative' of a given genus.

Acetobacter Genus. Gram type negative. Ovoid cells/rods, 0.6–0.8 × 1–4 μm. Non-motile or flagellate. Strictly aerobic. Opt. 25–30 °C. Chemoorganoheterotrophic. Respiratory. Many strains can oxidize ethanol to acetic acid/CO_2; used in vinegar production. Sugars probably metabolized mainly via the HMP and TCA pathways. GC% 51–65. Type species: *A. aceti.*

Acinetobacter Genus. Gram type negative. Rods, 0.9–1.6 × 1.5–2.5 μm. Non-motile. Strictly aerobic. Opt. usually 33–35 °C. Chemoorganoheterotrophic. Respiratory. Most strains can grow on minimal salts together with acetate, ethanol or lactate; few strains use glucose. Oxidase –ve. Found in soil, water; opportunist pathogens in man. GC% ca. 38–47. Type species: *A. calcoaceticus.*

Actinobacillus Genus. Gram negative. Rods/cocci, ca. 0.3–0.5 × 0.6–1.4 μm; non-motile. Facultatively anaerobic. Opt. ca. 37 °C. Chemoorganoheterotrophic. Respiratory and fermentative. Complex nutrients needed. No gas from the fermentation of glucose, lactose etc. Found in man, animals; can be pathogenic. GC% 40–43. Type species: *A. lignieresii.*

Actinomyces Genus. Gram type positive. Rods or filaments, often branched; non-motile. No spores. Typically anaerobic/microaerophilic. Opt. ca. 37 °C. Chemoorganoheterotrophic. Primarily fermentative. Carbohydrates fermented anaerogenically. Found in warm-blooded animals, e.g. in the mouth; can be pathogenic. GC% ca. 57–73. Type species: *A. bovis.*

Actinomycetales Order. Gram type positive. Most genera aerobic. Cocci, rods, mycelium (depending on genus); typically non-motile. Spores in many genera. Found in soil, composts, water etc.; some species symbiotic in plants, some pathogenic in man, other animals or plants. Many genera, including *Actinomyces, Arthrobacter, Corynebacterium, Mycobacterium* and *Streptomyces*.

Aeromonas Genus. Gram negative. Rods or coccobacilli, 0.3–1.0 × 1–3.5 μm; singly, pairs, chains, filaments. Some species motile (usually monotrichous); psychrotrophic species non-motile. Facultatively anaerobic. Chemoorganoheterotrophic. Respiratory and fermentative. Sugars, organic acids used as carbon sources. Oxidase +ve. Found in marine and fresh waters; *A. salmonicida* (opt. 22–25 °C) is parasitic/pathogenic in fish. GC% 57–63. Type species: *A. hydrophila*.

Agrobacterium Genus. Gram negative. Rods, 0.6–1 × 1.5–3 μm, capsulated; motile. Aerobic. Opt. 25–28 °C. Chemoorganoheterotrophic. Respiratory. Glucose metabolized mainly via Entner–Doudoroff and HMP pathways (section 6.2). Found in soil; most strains can induce tumours in plants, pathogenicity being plasmid-encoded. GC% 57–63. Type species *A. tumefaciens*.

Alcaligenes Genus (taxonomically unsettled). Gram negative. Rods, coccobacilli, cocci; motile. Aerobic. Chemoorganoheterotrophic, some strains chemolithotrophic. Respiratory. Acetate, lactate, amino acids etc. used as carbon sources. Oxidase +ve. Found in soil, water, vertebrates etc. GC% ca. 56–70. Type species: *A. faecalis*.

Alteromonas Genus. Gram negative. Rods, 0.7–1.5 × 1.8–3 μm, some pigmented (yellow, orange, violet etc.); monotrichously flagellate. Aerobic. Chemoorganoheterotrophic. Respiratory. Carbon sources include acetate, alcohols, amino acids, sugars. Found in marine waters. GC% ca. 38–50. Type species: *A. macleodii*.

Anabaena Genus. Gram type negative. Filamentous cyanobacteria (q.v.); trichome: spherical, ovoid or cylindrical cells. Gas vacuoles. Heterocysts. *A. flos-aquae* can form 'blooms' (section 10.1.1), and produce toxins (*anatoxins*).

Aphanizomenon Genus. Gram type negative. Filamentous cyanobacteria (q.v.); trichomes: individual cells cylindrical, end cells are often tapering colourless 'hair cells'. Gas vacuoles. Heterocysts. May form 'blooms' (section 10.1.1) in fresh and brackish waters; some strains produce toxins.

Aquaspirillum Genus. Gram negative. Cells: typically helical, rigid, 0.2–1.4 × 2–30 μm (longer in some species). Motile. Aerobic/facultatively anaerobic. Chemoorganoheterotrophic/chemolithoautotrophic. Respiratory. Carbon sources: e.g. amino acids, not usually carbohydrates. Some species (e.g. *A. peregrinum*) can fix nitrogen anaerobically. Typically oxidase +ve. Found in various freshwater habitats. GC% 49–66. Type species. *A. serpens*.

Archaebacteria Kingdom. Differ from Eubacteria (q.v.) (and from eukaryotic organisms) e.g. in nucleotide sequences in 16S rRNA (section 16.2.3), and in the chemistry of the cytoplasmic membrane and cell wall (which lacks peptidoglycan). Archaebacteria are often found in 'harsh' environments—many being e.g. extreme thermophiles or halophiles (sections 3.1.4 and 3.1.7). Genera include e.g. *Desulfurococcus, Halobacterium, Thermoproteus*.

Azotobacter Genus. Gram negative. Rods/coccobacilli/filaments; motile or non-motile. Cysts (section 4.3.3). Aerobic. Chemoorganoheterotrophic. Carbon sources: e.g. sugars, ethanol. Can fix nitrogen (section 10.3.2.1). Most strains are oxidase +ve. Found e.g. in fertile soils of near-neutral pH. GC% 63–68. Type species: *A. chroococcum*.

Bacillus Genus. Gram type positive. Rods, often 0.5–1.5 × 2–6 μm, typically motile. Endospores (section 4.3.1). Aerobic or facultatively anaerobic, depending on species. Respiratory or facultatively fermentative. Most species chemoorganoheterotrophic; many can grow on nutrient agar. Some species (e.g. *B. polymyxa*) can fix nitrogen (section 10.3.2.1). *B. schlegelii* can grow chemolithoautotrophically. Found e.g. as saprotrophs in soil and water; some species cause disease in man and other animals (including some insects). GC% 30-70. Type species: *B. subtilis*.

Bacteroides Genus. Gram negative. Rods or filaments (some pigmented), non-motile or motile. Anaerobic. Characteristically fermentative; some strains can carry out anaerobic respiration. Most species use sugars; others use peptones. Found e.g. in the alimentary tract in warm-blooded animals; some species are opportunist pathogens. GC% 28-61. Type species: *B. fragilis*. (As indicated e.g. by the GC% range, the genus is heterogeneous; some workers believe that it should contain only *B. fragilis* and closely-related organisms [Shah & Collins (1989) IJSB *39*, 85-87].)

Bartonella Genus. Gram negative. Rods, pleomorphic, 1-3 μm long, motile in culture. The sole species, *B. bacilliformis*, is the causal agent of Oroya fever (section 11.3.3).

Bdellovibrio Genus. Gram negative. Cells: vibrioid, 0.2-0.5 × 0.5-1.4 μm, each with one sheathed flagellum. Aerobic. Respiratory. Chemoorganoheterotrophic. Predatory: grow within the periplasmic space, and digest, other bacteria (e.g. *Aquaspirillum serpens, Escherichia coli, Pseudomonas* spp). Found e.g. in soil and sewage. GC% 33-51. Type species: *B. bacteriovorus*.

Beggiatoa Genus. Gram negative. Trichomes. Aerobic/microaerophilic/anaerobic. Respiratory. Typically chemoorganoheterotrophic; carbon sources: e.g. acetate, but hexoses (e.g. glucose) are not used. Found in various aquatic habitats.

Beijerinckia Genus. Gram negative. Rods, typically 0.5-1.5 × 1.7-4.5 μm, motile or non-motile. Aerobic. Respiratory. Chemoorganoheterotrophic; sugars (e.g. glucose) are used as carbon sources. Can fix nitrogen (section 10.3.2.1). Opt. 20-30 °C; no growth at 37 °C. Found e.g. in soil and on leaf surfaces. GC% 55-61. Type species: *B. indica*.

Bordetella Genus. Gram negative. Coccobacilli, approx. 0.2-0.5 × 0.5-2 μm, non-motile or motile. Aerobic. Respiratory. Carbon sources: e.g. amino acids; sugars not used. Enriched media are needed for culture. Found e.g. as parasites/pathogens of the mammalian respiratory tract. GC% 66-70. Type species: *B. pertussis*.

Borrelia Genus (order Spirochaetales—q.v. for basic details). Cells: about 0.2-0.5 × 3-20 μm. Anaerobic/microaerophilic. Some species can be grown in complex media. Found as parasites/pathogens in man and other animals. Type species: *B. anserina*.

Brucella Genus. Gram negative. Rods, coccobacilli or coccoid cells, about 0.5-0.7 × 0.6-1.5 μm; non-motile. Aerobic. Respiratory. Chemoorganoheterotrophic. Complex, enriched media are needed for culture. Most strains are oxidase +ve, urease −ve. Found, typically, as intracellular parasites/pathogens of animals, including man. GC% 55-58. Type species: *B. melitensis*.

Campylobacter Genus. Gram negative. Cells: spiral, typically 0.2-0.5 × 0.5-2 μm; motile, with a single unsheathed flagellum at one or both poles. Microaerophilic, needing 3-5% CO_2 for growth. Respiratory. Chemoorganoheterotrophic; carbon sources: amino acids or TCA cycle intermediates (Fig. 5.10) but not carbohydrates. Oxidase +ve. Found e.g. in the reproductive and intestinal tracts in man and other animals. GC% 30-38. Type species: *C. fetus*. (*C. pylori* has been transferred to a new genus, *Helicobacter* [Goodwin *et al*. (1989) IJSB *39*, 397-405].)

Caulobacter Genus. Gram negative. Complex cell cycle (section 4.1). Some strains are pigmented. Aerobic. Respiratory. Chemoorganoheterotrophic. Found in certain soils and waters. GC% 64-67.

Cellulomonas Genus. Gram type positive. Rods/filaments/coccoid cells, non-motile or motile. Aerobic/facultatively anaerobic. Respiratory and fermentative. Chemoorganoheterotrophic; starch and cellulose are attacked. Found e.g. in soil. GC% 71–77. Type species: *C. flavigena*.

Chlamydia Genus. Gram negative. Cells: vary according to stage in the developmental cycle, but are non-motile, coccoid, 0.2–1.5 μm in diameter, and pleomorphic. Obligate intracellular parasites/pathogens in man and other animals; have been cultured in laboratory animals, in the yolk sac of chick embryos, and in cell cultures. GC% 41–44. Types species: *C. trachomatis*.

Chlorobium Genus. Gram negative. Rods or vibrios, about 1–2 μm long, non-motile. Obligately anaerobic. Primarily photolithoautotrophic (electron donors: e.g. sulphide). Chlorophyll occurs in *chlorosomes* (section 2.2.7). Found e.g. in sulphide-rich mud.

Clostridium Genus. Gram type positive. Cells: typically rods, about 0.3–1.9× 2–10 μm, motile or non-motile. Endospores (section 4.3.1). Obligately anaerobic (or, in a few cases, aerotolerant). Chemoorganoheterotrophic. Typically fermentative, though some strains (of e.g. *C. perfringens*) can carry out nitrate respiration (section 5.1.1.2). Growth often poor in/on basal media. Found e.g. in soil and in the intestines of man and other animals; some species pathogenic. GC% 22–55. Type species: *C. butyricum*.

coliform In general: any Gram-negative, non-sporing, facultatively anaerobic bacillus which can ferment lactose within 48 hours with the formation of acid and gas at 37 °C—or (as defined in the USA) with gas formation at 35 °C. In the UK, water bacteriologists require that a coliform also be (i) oxidase –ve, and (ii) able to grow aerobically and anaerobically in the presence of certain surfactants (e.g. bile salts); for dairy microbiologists, lactose fermentation should occur within the range 30–35 °C. *Escherichia coli* is a typical coliform.

Corynebacterium Genus. Gram positive. Rods, often curved/pleomorphic, non-motile. Facultatively anaerobic. Chemoorganoheterotrophic. Respiratory and fermentative. Found e.g. in soil and vegetable matter; some species parasitic/ pathogenic in man and other animals. GC% 51–59. Type species: *C. diphtheriae*.

Coxiella Genus. Gram negative. Rods (highly pleomorphic), 0.2–0.4× 0.4–1 μm, non-motile. Endospores (section 4.3.1). The sole species, *C. burnetii*, is an obligate intracellular parasite/pathogen in vertebrates and arthropods; it undergoes a developmental cycle. GC% about 43.

cyanobacteria ('blue-green algae') Non-taxonomic category. Photosynthetic bacteria which differ from most other phototrophic prokaryotes (i) in having chlorophyll *a*, and (ii) in carrying out *oxygenic* photosynthesis (section 5.2.1.1) (cf. *Prochloron*). Cells: Gram type negative, single or e.g. in trichomes (according to species). No flagella; some exhibit gliding motility (section 2.2.15.1). Some have gas vacuoles (section 2.2.5). Depending on e.g. pigments, cells may appear blue-green, yellowish, red, purple or almost black etc. Some species form akinetes, heterocysts and/or hormogonia (section 4.4).

Typically photolithoautotrophic, fixing CO_2 via the Calvin cycle (section 6.1.1), and respiratory, using oxygen as terminal electron acceptor. Some can grow as chemoorganoheterotrophs, carrying out e.g. anaerobic respiration or even fermentation. Some can carry out facultative *anoxygenic* photosynthesis, using photosystem I (Fig. 5.11) with e.g. sulphide as electron donor.

Found in a wide range of aquatic and terrestrial habitats; some form 'blooms' (section 10.1.1).

Desulfomonas Genus. Gram negative. Rods, non-motile. Anaerobic. Respiratory: sulphate respiration (section 5.1.1.2) using e.g. pyruvate as electron donor. Found e.g. in the human intestine. GC% 66–67. Type species: *D. pigra*.

Desulfurococcus Genus. Archaebacteria (q.v.). Cocci, about 1 μm diameter, motile or non-motile. Anaerobic, thermophilic, chemolithoheterotrophic. Respiratory (sulphur respiration: section 5.1.1.2). GC% 51. Found e.g. in Icelandic solfataras.

Enterobacteriaceae Family. Gram-negative, non-sporing, facultatively anaerobic bacilli, typically $0.3-1 \times 1-6\,\mu m$; motile (most peritrichously flagellate) or non-motile. Cells occur singly or in pairs. Chemoorganoheterotrophic; typically grow well in/on basal media (section 14.2.1). Carbon sources include sugars. Respiratory *and* fermentative. Oxidase −ve. All except a few strains are catalase +ve. Found e.g. as parasites, pathogens or commensals in man and other animals, and as saprotrophs in soil and water.

Genera (and species) are differentiated e.g. by biochemical tests—particularly IMViC tests (section 16.1.2.5), the urease test (section 16.1.2.7), and the decarboxylase tests (section 16.1.2.8). Genera include e.g. *Citrobacter, Enterobacter, Erwinia, Escherichia, Klebsiella, Proteus, Salmonella, Serratia, Shigella* and *Yersinia*.

Erwinia Genus (family Enterobacteriaceae—q.v. for basic details). Saprotrophic, or pathogenic in plants and animals. Typically motile. Acid (little/no gas) from sugars. Opt. 27–30 °C. GC% 50–58. Type species: *E. amylovora*.

Escherichia Genus (family Enterobacteriaceae—q.v. for basic details). The following refers to *E. coli*. Cells: single or in pairs, typically motile (peritrichously flagellate) and fimbriate (section 2.2.14.2, see also items on Plates 1, 2 and 4). Opt. 37 °C. Respiratory under aerobic conditions; fermentation (section 5.1.1.1) or e.g. nitrate respiration (section 5.1.1.2) carried out anaerobically. Glucose is fermented (usually with gas) via the mixed acid fermentation (Fig. 5.5). *Typical* reactions as follows. IMViC tests (section 16.1.2.5): +, +, −, −; citrate +ve in strains containing the Cit plasmid (section 7.1); urease −ve; H_2S −ve; lactose +ve (acid and gas). Found e.g. as part of the normal microflora of the intestine in man and other animals; some strains can be pathogenic (sections 11.2.1 and 11.3.3). GC% 48–52. Type species: *E. coli*.

Eubacteria Kingdom. Includes most or all bacteria not classified in the Archaebacteria (q.v.)—e.g. most Gram-positive bacteria, all the cyanobacteria and the anoxygenic photosynthetic bacteria, all enterobacteria and pseudomonads, and the mycoplasmas. Eubacteria differ from members of the Archaebacteria e.g. in their 16S rRNA (section 16.2.3) and in the chemistry of their cytoplasmic membrane (section 2.2.8) and cell wall (section 2.2.9).

The medically important bacteria—and those species most likely to be encountered in an introductory course in bacteriology—are eubacteria.

Francisella Genus. Gram negative. Cocci, coccobacilli or rods (depending on species and conditions), non-motile. Aerobic. Chemoorganoheterotrophic; carbohydrates metabolized slowly, without gas. Opt. 37°C. Oxidase −ve. Catalase weakly +ve. Found as parasites/pathogens of man and other animals. GC% 33–36. Type species: *F. tularensis* (formerly *Pasteurella tularensis*).

Gardnerella Genus. Gram type negative (?). Rods (pleomorphic), about $0.5 \times 1.5-2.5\,\mu m$. Obligately anaerobic and facultatively anaerobic strains. Chemoorganoheterotrophic; growth occurs only on enriched media. Oxidase −ve. Catalase −ve. Opt. 35–37 °C. Found in the human genital/urinary tract; can be pathogenic. GC% about 42–44. Type species: *G. vaginalis* (formerly *Haemophilus vaginalis*).

Haemophilus Genus. Gram negative. Rods/coccobacilli (pleomorphic), often about $0.4 \times 1-2\,\mu m$, or filaments; non-motile. Facultatively anaerobic. Respiratory and fermentative. Chemoorganoheterotrophic; growth occurs on enriched media, e.g. chocolate agar (section 14.2.1). Typically, glucose, but not lactose, is fermented. Opt. 35–37 °C. Found as parasites/pathogens in man and other animals. GC% 37–44. Type species: *H. influenzae*.

Halobacterium Genus. Archaebacteria (q.v.). Rods or filaments, motile or non-motile. Gas vacuoles (section 2.2.5) common. Extremely halophilic. Facultatively anaerobic. Aerobic metabolism: chemoorganoheterotrophic and respiratory, with e.g. amino acids or carbohydrates as carbon sources. Oxidase +ve. Some strains obtain energy from a purple membrane (section 5.2.2). Found e.g. in evapourated brines, salted fish etc. GC% 66–68. Type species: *H. salinarium* (formerly *H. halobium*).

Helicobacter Genus. Gram negative. Cells: helical, motile with several sheathed flagella (section 2.2.14.1, Plate 1: top, left). Microaerophilic. Chemoorgano-heterotrophic. Urease +ve. *H. pylori* is associated with disease of the human gastrointestinal tract; it was originally classified in the genus *Campylobacter* (q.v.).

Klebsiella Genus (family Enterobacteriaceae—q.v. for basic details). Cells: single, pairs, short chains; capsulated. Non-motile. Often MR –ve, VP +ve. Found e.g. in soil, water, and as parasites/pathogens in man and other animals. GC% 53–58. Type species: *K. pneumoniae*.

Kurthia Genus. Gram positive. Rods or filaments; rods are peritrichously flagellate. Aerobic. Respiratory. Chemoorganoheterotrophic; amino acids, alcohols, fatty acids used as carbon sources. Found e.g. on meat and meat products. GC% about 36–38. Type species: *K. zopfii*.

Lactobacillus Genus. Gram positive. Rods or coccobacilli, singly or in chains; typically non-motile. Anaerobic, microaerophilic or facultatively aerobic; usually catalase –ve. Chemoorganoheterotrophic, using e.g. sugars as carbon sources. Characteristically fermentative, lactic acid being formed from glucose by homo-lactic fermentation (section 5.1.1.1) or by a heterolactic fermentation in which mixed products, including lactic acid, are formed. Found e.g. on vegetation, as part of the natural microflora in man, and in various fermented food products. GC% about 32–53. Type species: *L. delbrueckii*.

Legionella Genus. Gram negative. Rods/filaments, 0.3–0.9 × 2–>20 μm; motile. Aerobic. Chemoorganoheterotrophic, using amino acids (non-fermentatively) for carbon and energy; growth occurs e.g. on blood agar containing L-cysteine and iron. Opt. 35–37 °C. Found e.g. in various aquatic habitats (such as thermally polluted streams); most/all species can be pathogenic for man. Type species: *L. pneumophila*.

Leuconostoc Genus. Gram positive. Cells: coccoid, about 1 μm in diameter, in pairs or chains; non-motile. Facultatively anaerobic. Fermentative and respiratory; anaerobically, glucose is fermented mainly to lactic acid, ethanol and CO_2. Found e.g. in various dairy products and fermented drinks. GC% about 38–44. Type species: *L. mesenteroides*.

Listeria Genus. Gram positive. Rods or coccobacilli, about 0.5 × 0.5–2 μm; motile when grown at 20–25 °C, non-motile at 37 °C. Aerobic, facultatively anaerobic. Chemoorganoheterotrophic. Catalase +ve. Oxidase +ve. Sugars are fermented (acid, no gas). Found in soil, decaying vegetation, and as pathogens in man and other animals. Type species: *L. monocytogenes*.

methanogens Non-taxonomic category. Includes all bacteria able to produce methane (section 5.1.2); all are obligately anaerobic archaebacteria which occur e.g. in mud and in the rumen. Genera include *Methanobacterium, Methanobrevibacter, Methanococcus* and *Methanolobus*.

Methylococcus Genus. Gram negative. Cocci, about 1 μm in diameter; non-motile. Aerobic/microaerophilic. Obligately methylotrophic (section 6.4); methane can be used as sole source of carbon and energy. Found e.g. in mud, soil. GC% about 63. Type species: *M. capsulatus*.

Mycobacterium Genus. Gram positive. Rods, 0.2–0.8 × 1–10 μm, coccoid forms, branched rods or fragile filaments; some strains pigmented. Non-motile. Acid-fast (section 14.9.2) during at least some stage of growth. Microaerophilic. Respiratory. Typically chemoorganoheterotrophic, though some strains may be chemolithotrophic; typically not nutritionally fastidious, though growth in at least some can be stimulated e.g. by serum or egg-yolk. Found e.g. as free-living saprotrophs in soil and water, or on plants, and as parasites/pathogens of man and other animals. GC% about 62–70. Type species: *M. tuberculosis*.

Mycoplasma Genus. Cells: pleomorphic, ranging from coccoid (about 0.3–0.8 µm in diameter) to branched filamentous forms; some capable of gliding motility (section 2.2.15.1). No cell wall. Facultatively or obligately anaerobic. Chemoorganoheterotrophic; growth occurs on complex media, and all species need cholesterol or related sterols. Catalase –ve. Found as parasites/pathogens e.g. in the respiratory and urogenital tracts in man and other animals. GC% about 23–40. Type species: *M. mycoides*.

Neisseria Genus. Gram type negative. Typically cocci, 0.6–1 µm in diameter; non-motile. Aerobic. Chemoorganoheterotrophic; some species need enriched media (e.g. chocolate agar). Oxidase +ve. Found e.g. as parasites/pathogens of man and other animals. GC% about 46–54. Type species: *N. gonorrhoeae*.

Nitrobacter Genus. Gram negative. Rods, about 0.6–0.8 × 1–2 µm; usually non-motile. Reproduce by budding (section 3.2.2). Obligately aerobic. Some strains obligately chemolithoautotrophic (nitrifying bacteria: section 5.1.2; Fig. 10.2), others facultatively chemoorganoheterotrophic. Opt. 25–30 °C. Found e.g. in soil. GC% about 61. Type species: *N. winogradskyi*.

Nitrosococcus Genus. Gram negative. Cocci, about 1.5 µm in diameter; motile or non-motile. Obligately aerobic. Obligate chemolithoautotrophs, oxidizing ammonia to nitrite (section 5.1.2; Fig. 10.2). Found e.g. in soil.

Oscillatoria Genus. Gram negative. Filamentous cyanobacteria (q.v.); trichomes: motile, composed of flattened, disc-shaped cells. Gas vacuoles. Hormogonia. Found in various aquatic and terrestrial habitats. GC% 40–50.

Pelodictyon Genus. Gram negative. Rods/coccoid forms which may form chains/three-dimensional networks; non-motile. Gas vacuoles. Anaerobic. Phototrophic. Found e.g. in sulphide-rich mud.

Prochloron Genus. Gram type negative (?). Cocci, 6–25 µm in diameter. Cells contain chlorophylls *a* and *b* and carry out oxygenic photosynthesis. Taxonomy uncertain: the only species, *P. didemni* (found on warm-water sea-squirts), resembles cyanobacteria in having chlorophyll *a*, but differs in having chlorophyll *b* and in lacking certain typical cyanobacterial pigments. Possibly related to ancestral chloroplasts; a puzzle.

Propionibacterium Genus. Gram positive. Pleomorphic branched/unbranched rods or coccoid forms; non-motile. Anaerobic. Chemoorganoheterotrophic. Fermentative: hexoses (e.g. glucose) or lactate fermented mainly to propionic acid. Growth occurs e.g. on yeast extract–lactate–peptone media. Found e.g. in dairy products. GC% about 57–67. Type species: *P. freudenreichii*.

Proteus Genus (family Enterobacteriaceae—q.v. for basic details). Motile, often swarming (section 4.2). Typically H₂S +ve, urease +ve. Growth requires nicotinic acid. Found e.g. in soil, polluted waters, and the mammalian intestine; some (e.g. *P. mirabilis*) can be pathogenic. GC% 38–41. Type species: *P. vulgaris*.

Pseudomonas Genus. Gram negative. Rods, 0.5–1 × 1.5–5 µm; most species have one/several unsheathed, typically polar flagella per cell, though *P. mallei* is non-motile (i.e. it lacks flagella), and some species have sheathed flagella (section 2.2.14.1). Aerobic or facultatively anaerobic. Respiratory; many species can carry out nitrate respiration (section 5.1.1.2). Typically chemoorganoheterotrophic and nutritionally highly versatile; many strains will grow on inorganic salts with an organic carbon source, while some can grow chemolithoautotrophically. Catalase +ve. Commonly oxidase +ve. Found e.g. in soil and water, and as pathogens in man, other animals, and plants. GC% 58–70. Type species: *P. aeruginosa*.

Pyrodictium Genus. Archaebacteria (q.v.). The organisms grow as a network of filaments associated with 'discs', each disc being 0.3-2.5 μm in diameter. Anaerobic. Energy obtained by the metabolism of elemental sulphur. Chemolithoautotrophic. Thermophilic (section 3.1.4). Halotolerant. Found in an underwater volcanic region.

Rhizobium Genus. Gram negative. Rods, 0.5-0.9× 1.2-3 μm; motile. Aerobic. Chemoorganoheterotrophic; carbon sources include sugars. Found e.g. in soil and in root nodules (section 10.2.4.1). GC% 59-64. Type species: *R. leguminosarum*.

Rickettsia Genus. Gram negative. Rods, 0.3-0.6× 0.8-2 μm; non-motile. Obligate intracellular parasites/pathogens in vertebrates (including man) and arthropods (ticks, mites etc.). Apparently respiratory, with glutamate as the main energy subtrate; glucose is not used. Opt. 32-35 °C. GC% about 29-33. Type species: *R. prowazekii*.

Ruminococcus Genus. Gram type positive. Cocci, about 1 μm in diameter, in pairs or chains. Anaerobic. Chemoorganoheterotrophic; typically heterofermentative, forming e.g. acetic and formic acids from carbohydrates. Many strains can use cellulose. Found in the rumen. GC% about 40-45. Type species: *R. flavefaciens*.

Salmonella Genus (family Enterobacteriaceae—q.v. for basic details). Typically motile. *Typical* reactions as follows. IMViC tests (section 16.1.2.5): -, +, -, +; glucose (acid and gas at 37 °C) +ve; lactose usually -ve (but the ability to ferment lactose can be plasmid-encoded); H_2S +ve; urease -ve; lysine and ornithine decarboxylases (section 16.1.2.8) +ve. Salmonellae can grow on basal media and e.g. on MacConkey's agar and DCA (section 14.2.1); enrichment media include e.g. selenite broth (Table 14.1). Found e.g. as pathogens in man and other animals. GC% 50-52. Type species: *S. choleraesuis*.

Unlike most bacteria, the salmonellae are commonly identified and named as *serotypes* (section 16.1.4.2) rather than as species. In the *Kauffmann-White classification scheme* there are about 2000 named serotypes; each serotype is defined by its O and H antigens (section 16.1.4.2) and, in some serotypes, by the *Vi antigen*: an antigen in a polysaccharide microcapsule (section 2.2.11) associated with virulence for particular host(s). Each serotype is given an antigenic formula which lists, in order, the organism's O, Vi (if present) and H antigens; in many serotypes the H antigens can switch, owing to *phase variation* (see Fig. 8.3c), so that the formula of such a serotype includes two alternative H antigens (or two alternative *sets* of H antigens). For example, the antigenic formula of *S. typhimurium* is 1,4,[5],12:i:1,2. This means: O antigens 1, 4, 5 ([ ] indicates variable presence) and 12, phase 1 H antigen 'i', and phase 2 H antigens 1 and 2; O antigen 1 is underlined to show that that antigen is present as a result of *phage conversion* (section 9.4).

Serratia Genus (family Enterobacteriaceae—q.v. for basic details). Usually motile. Some strains form a red pigment, *prodigiosin*. *Typical* reactions: MR -ve; VP +ve (at 30 °C, but may be -ve at 37 °C); citrate +ve; lactose +ve or -ve, according to species. Glucose is fermented by the Entner-Doudoroff pathway (Fig. 6.2). Found e.g. in soil and water, on plants, and in man and other animals. GC% 52-60. Type species: *S. marcescens*.

Shigella Genus (family Enterobacteriaceae—q.v. for basic details). Non-motile. Sugars fermented usually without gas. MR +ve; VP -ve; citrate -ve; H_2S -ve; lysine decarboxylase -ve. Found e.g. as intestinal pathogens of man and other primates. GC% 49-53. Type species: *S. dysenteriae*.

Simonsiella Genus. Gram negative. Flat, multicellular filaments, the outer face of each terminal cell being rounded (Plate 1: top, right). Gliding motility. Chemoorganoheterotrophic. Found e.g. in the mouth (human and animal).

Spirochaetales Order. Helical, flexible, Gram type negative. Each cell resembles a Gram-negative bacillus in which one or more periplasmic flagella (= 'axial fibrils') (sections 2.2.14.1 and 2.2.15.1) arise at *each* end of the cell and wind around the protoplast (= 'protoplasmic cylinder'); cells are 0.1–3 × 5–250 μm, according to species. The spirochaetes include free-living and pathogenic species, anaerobic and aerobic species. Chemoorganoheterotrophic. Respiratory and/or fermentative. Genera include *Borrelia* and *Treponema*.

Staphylococcus Genus. Gram positive. Cocci, about 1 μm in diameter, often in clusters, some containing orange or yellow carotenoid pigments; non-motile. Facultatively anaerobic. Chemoorganoheterotrophic. Carbon sources include various sugars. Commonly halotolerant (section 3.1.7). Catalase +ve. The staphylococci are divided into coagulase +ve and coagulase –ve strains (section 16.1.2.2), the former including *S. aureus* and *S. intermedius*, the latter including *S. epidermidis* (formerly *S. albus*). Found e.g. as commensals and pathogens of man and other animals. GC% about 30–39. Type species: *S. aureus*.

Streptococcus Genus. Gram positive. Cocci, or coccoid forms, typically 1 μm in diameter, often in pairs or chains; usually non-motile. Capsulation common. Facultatively anaerobic. Catalase –ve, though a pseudocatalase (section 16.1.1.6) may occur in some strains. Chemoorganoheterotrophic. Typically fermentative, sugars being metabolized usually without gas (glucose often yielding lactic acid); aerobically, and in certain media, some strains can carry out respiratory metabolism. Found e.g. as commensals and pathogens in man, and in various dairy products. Type species: *S. pyogenes*.

(Proposals have been made to transfer some species of *Streptococcus* to other genera—e.g. *S. avium*, *S. durans*, *S. faecalis*, *S. faecium* and *S. gallinarum* to *Enterococcus; S. cremoris* and *S. lactis* to *Lactococcus*.)

Streptomyces Genus (order Actinomycetales—q.v.). Gram positive. Mycelium (section 2.1.1), part of which fragments to form chains of spores (section 4.3.2; Fig. 4.3). Aerobic. Respiratory. Chemoorganoheterotrophic; carbon sources include glucose, lactate and starch. Antibiotics formed by *Streptomyces* species include chloramphenicol (section 15.4.4) and streptomycin (section 15.4.2). Found e.g. in soil and as pathogens of plants. GC% 69–78. Type species: *S. albus*.

Sulfolobus Genus. Archaebacteria (q.v.). Cocci, coccoid or irregularly-shaped cells in which the cell wall consists only of an S layer (section 2.2.12). Thermophilic (growth occurs between 50 and 90 °C). Acidophilic (section 3.1.5). Aerobic and facultatively anaerobic. Energy is obtained by the (respiratory, aerobic) oxidation of sulphur (or Fe^{2+}) and/or by sulphur respiration (section 5.1.1.2) in which elemental sulphur is used as terminal electron acceptor. Obligately heterotrophic or facultatively autotrophic. Found e.g. in certain hot springs.

Thermoproteus Genus. Archaebacteria (q.v.). Rods, filaments, about 0.5 × 1–80 μm. Anaerobic. Energy obtained by sulphur respiration (see also *Sulfolobus*). Thermophilic. Autotrophic and/or heterotrophic (carbon sources include glucose, ethanol, formate). Found e.g. in Icelandic solfataras.

Thiobacillus Genus. Gram negative. Rods, about 0.5 × 1–3 μm; typically motile. Obligately aerobic or (some) facultatively anaerobic. Respiratory; energy commonly obtained by the oxidation of sulphur and/or reduced sulphur compounds. Obligately or facultatively chemolithoautotrophic. Found e.g. in soil, mud, hot springs. GC% about 50–68. Type species: *T. thioparus*.

Treponema Genus (order Spirochaetales—q.v. for basic details). Cells: 0.1–0.4 × 5–20 μm. Anaerobic or microaerophilic. Some species can be grown in complex media; others (including *T. pallidum*) cannot, and are grown e.g. intratesticularly in rabbits. Found e.g. as parasites/pathogens in man and other animals. GC% 25–54. Type species: *T. pallidum*.

Vibrio Genus. Gram negative. Rods, curved (vibrios) or straight, 0.5–0.8 × 1.4–2.6 μm; motile, flagella typically sheathed (section 2.2.14.1). Facultatively anaerobic.

Typically oxidase +ve. Chemoorganoheterotrophic; glucose is fermented by the mixed acid fermentation (Fig. 5.5), usually without gas. All species can grow at 20 °C, most at 30–35 °C, and some at 40 °C. Some species tolerate high pH (e.g. *V. cholerae* can grow at pH 10). Found e.g. in various aquatic habitats (freshwater, estuarine and marine) and as pathogens in man, fish and shellfish. GC% 38–51. Type species: *V. cholerae*.

Xanthomonas Genus. Gram negative. Rods, 0.4–0.7 × 0.7–1.8 μm, typically containing yellow pigment(s); some strains form extracellular slime. Motile. Aerobic. Chemoorganoheterotrophic. In strains of X. *campestris*, glucose is metabolized e.g. via the Entner–Doudoroff pathway (Fig. 6.2). Oxidase –ve (or weakly +ve). Catalase +ve. Found e.g. as pathogens in plants. GC% 63–71. Type species: X. *campestris*.

Yersinia Genus (family Enterobacteriaceae—q.v. for basic details). Cells: 0.5–0.8 × 1–3 μm. Most species are motile < 30 °C, non-motile at 37 °C; Y. *pestis* is non-motile. Growth occurs on basal media. VP – ve at 37 °C (+ve in some species at 25 °C); MR +ve; acid (little/no gas) from glucose; lactose typically – ve. Opt. growth temperature: 28–29 °C. Found e.g. as parasites/pathogens in man and animals. GC% 46–50. Type species: Y. *pestis*.

Index

741011